Fundamentals of the Immune System

FUNDAMENTALS OF THE IMMUNE SYSTEM

Cheryl Natividad

Delve
PUBLISHING
www.delvepublishing.com

Fundamentals of the Immune System

Cheryl Natividad

Delve Publishing
224 Shoreacres Road
Burlington, ON L7L 2H2
Canada
www.delvepublishing.com
Email: orders@arclereducation.com

ISBN: 978-1-77469-398-8 (Hardcover)

ABOUT THE AUTHOR

Cheryl Agdaca- Natividad graduated with a BS Biology degree (magna cum laude) at the Saint Louis University, Baguio City in the Philippines. She took the licensure examination for secondary teachers in 2003 where she ranked no. 10 among the examinees. She earned her masters degree in Genetics at the premier state university, University of the Philippines Los Baños. She taught in the same university from 2006-2017 handling lecture and laboratory classes in general biology, cell biology, genetics, and molecular genetics. She authored books on various fields of genetics such as veterinary genetics, plant genetics, and epigenetics.

TABLE OF CONTENTS

LIST OF FIGURES

Figure 2.5. (a) A small, unstimulated lymphocyte has little amount of cytoplasm with purple-staining nucleus. (b) A large, stimulated one becomes granular with more cytoplasm. (Source: a-Nicolas Grandjean, Creative Commons License; b-SpicyMilkBoy, Creative Commons License)

Figure 2.6. Monocytes are the largest among the leukocytes with bilobed nuclei. (Source: Left-Graham Beards, Creative Commons License; Right-Bruce Blaus & Blausen.com staff, Creative Commons License)

Figure 2.7. Monocytes originate from progenitor cells (i.e., common myeloid progenitor cells, granulocyte-macrophage progenitor cells, common macrophage precursor cells, committed monocyte progenitor cells) in lymphoid organs like the bone marrow. In the peripheral blood, the monocytes have the capacity to differentiate into three subsets that have distinct functions and phenotypes in terms of CD14 and CD16 expression. Additionally, monocytes can also undergo transendothelial migration and subsequently differentiate into macrophages, which are activated and polarized to acquire pro-inflammatory (M1), anti-inflammatory (M2), or the regulatory (Mreg) phenotype. (Source: Thierry P.P. van den Bosch *et al*., Creative Commons License)

Figure 2.8. Macrophages are phagocytes. In this case, a macrophage engulfed a fungal pathogen (pale yellow circle). (Source: Carolina Coelho, Creative Commons License)

Figure 2.9. Tissue-resident macrophages are produced at different stages of development and arise from three different sources. In the early stage of embryogenesis, myeloid progenitors from the yolk sac give rise to microglia in the brain, Kupffer cells in the liver, and Langerhans cells in the skin. When hematopoiesis starts to occur in the liver of the fetus, fetal liver-derived monocytes differentiate and contribute to the pool of Langerhans cells in the skin and macrophages in the gut lamina propria. They also seed the lung with pre-alveolar macrophages before birth. After birth, these rapidly differentiate into long-lived alveolar macrophages. Unlike the other tissue macrophages that proliferate after birth, the lamina propria macrophages are continuously renewed through the differentiation of bone marrow-derived monocytes. Monocytes can also be recruited to sites of infection or injury and differentiated into inflammatory macrophages, monocyte-derived dendritic cells (Mo-DCs), or myeloid-derived suppressor cells (MDSCs). (Source: Ismé De Kleer *et al*., Creative Commons License)

Figure 2.10. A follicular dendritic cell (FDC) has numerous cell processes or dendrites that interconnect with one another and with dendrites of adjacent FDCs. These form a network that can trap antigens and initiate an antibody response. (Source: Askazal, Creative Commons License)

Figure 2.11. There are a series of three progenitor cells, i.e., granulocyte, macrophage, and dendritic cell progenitors (GMDP); macrophage, dendritic cell progenitor (MDP); and common dendritic cell progenitor (CDP) that differentiate in a step-wise manner into classical (cDC) and plasmacytoid dendritic cell (pDC). The differentiation process is influenced by certain growth factors such as fms-like tyrosine kinase 3 ligand (Flt3L), granulocyte-macrophage colony-stimulating factor (GM-CSF), and macrophage colony-stimulating factor (M-CSF). (Source: Andrés Castell-Rodríguez *et al.*, Creative Commons License)

Figure 2.12. The two types of conventional dendritic cells, cDC1 andcDC2, may be located in the blood, skin, mucous membranes of the digestive and respiratory tracts as well as in lymphoid organs such as the spleen. (Source: Andrés Castell-Rodríguez *et al.*, Creative Commons License)

Figure 2.13. Chemokines (yellow circles) are secreted from the blood vessel wall or the underlying tissue in response to inflammatory signals like infection or tissue damage. These are then transported to the luminal surface of the endothelial cells where they bind to their receptors or to chemokine-interacting partners such as glycosaminoglycan (GAG). Thus, the chemokines form an immobilized gradient that drives the migration of leukocytes towards the inflammatory site. (Source: Amanda E.I. Proudfoot *et al.*, Creative Commons License)

Figure 2.14. Interferon (IFN) signaling influences the recognition of intracellular pathogens, such as bacteria and influenza A virus (IAV). Type I IFN activates the transcription factor interferon regulatory factor 1 (IRF1), which initiates the expression of guanylate-binding proteins (GBPs) and interferon response gene B10 (IRGB10). These gene products permeabilize the membrane of Gram-negative bacteria resulting in the release of bacterial DNA and lipopolysaccharide (LPS). AIM2 inflammasome detects bacterial DNA while caspase 11 (casp11) directly interacts with LPS. Type I IFN signaling also mediates the increased expression of Z-DNA-binding protein 1 (ZBP1), which recognizes the IAV proteins and triggers inflammasome activation and induces apoptosis, necrosis, and pyroptosis of IAV-infected cells. (Source: Nataša Kopitar-Jerala, Creative Commons License)

Figure 2.15. Tumor necrosis factor α (TNF) exhibits immunosuppressive function via its intrinsic negative effect on conventional T cells (Tconvs) or its stimulatory effect on suppressive cells, such as the myeloid-derived suppressor cell (MDSC) or regulatory T cell (Treg). (Source: Benoît L. Salomon, Creative Commons License)

Figure 2.16. A pattern recognition receptor (PRR) recognizes and binds certain structures common to pathogens such as lipopolysaccharide (LPS) in the membrane of Gram-negative bacteria. This activates a signaling pathway that upregulates the expression of cytokine genes, eventually resulting in the secretion of cytokines. (Source: Immcarle105, Creative Commons License)

Figure 2.17. A typical Toll-like receptor (TLR) has three important regions, namely the horseshoe-shaped leucine-rich repeat (LRR) motif that protrudes into the extracellular side, the transmembrane helix, and the intracellular Toll/IL-1 receptor (TIR) domain. TLRs dimerize when activated forming homodimers or in this case, a heterodimer (i.e., between TLR1 and TLR2). (Source: Wei Gao *et al.*, Creative Commons License)

Figure 2.18. Toll-like receptors (TLRs) are transmembrane receptors while nucleotide-binding oligomerization domain-leucin rich repeats-containing receptors (NLRs) are cytosolic ones. The different types of TLRs and NLRs (i.e., NOD1, NOD2, and NALP3 inflammasome) recognize certain pathogen associated molecular patterns (PAMPs) and damage-associated molecular patterns (DAMPs), activating signaling pathways. The TLR1/TLR2 and TLR6/TLR2 heterodimers, TLR4, and TLR3 trigger the TIR-domain-containing adapter-inducing interferon-β (TRIF) pathway causing the activation of interferon regulatory factor 3 (IRF3) and NF-κB. TLR5, TLR9, and TLR7 stimulate the myeloid differentiation primary-response gene 88 (MyD88) pathway activating NF-κB, IRF7, and mitogen-activated protein kinase (MAPK). NOD1 and NOD2 activate NF-κB, while NACHT, LRR, PYD domains-containing protein 3 (NALP3) inflammasome recruits and activates the caspase-1 pathway. (Source: Ji-Hyun Jang *et al.*, Creative Commons License)

Figure 2.19. Nucleotide-binding oligomerization domain leucine-rich repeats containing receptors (NLRs) have common domains, namely the nucleotide-binding oligomerization domain (NOD) at the middle portion and the leucine-rich repeats (LRR) domain at the C-terminal region (green). Human NLRs are classified based on their domains at the N-terminus. NLRA has an acidic transactivation domain (AD), NLRB has a baculovirus inhibitor of apoptosis protein repeat (BIR), NLRC has a caspase-recruitment and activation domain (CARD), and NLRP has a pyrin domain (PYD). (Source: Yifei Zhong *et al.*, Creative Commons License)

Figure 2.20. The binding of NOD2 to the bacterial peptidoglycan derivative muramyl dipeptide (MDP) induces it to form an oligomer with another NOD2. The oligomer interacts with receptor-interacting serine/threonine protein kinase 2 (RIPK2) resulting in the recruitment of E3 ubiquitin ligases like cellular inhibitor of apoptosis 1 (cIAP1), cIAP2, and X-linked inhibitor of apoptosis (XIAP). The ubiquitinated RIPK2 serves to engage and activate TAK1 and IKK to induce the NF-κB and MAPK signaling pathways. The NOD receptors are also thought to be activated by single-stranded RNA (ssRNA) viruses so that they interact with mitochondrial antiviral-signaling protein (MAVS) and activate the IRF3 pathway. They can also interact with NLRP via their CARD domains to activate the inflammasome. (Source: Yifei Zhong *et al.*, Creative Commons License)

Figure 2.21. There are three main classes of NLR inflammasomes. The NLRP1 inflammasome is made up of NLRP1, the adaptor apoptosis-associated speck-like protein containing a CARD domain (ASC), caspase-1, and caspase-5. It is possibly activated by the anthrax lethal toxin, bacterial muramyl dipeptide (MDP), and decreased cytosolic ATP. NLRP3 is stimulated by a wide variety of PAMPs and DAMPs. NLRC4 inflammasome responds to bacterial flagellin and type III secretion system (T3SS) proteins. (Source: Yifei Zhong *et al.*, Creative Commons License)

Figure 2.22. During viral infections, the NLRP3 inflammasome is activated through a two-signal mechanism. The priming signal is generated by the activation of pattern recognition receptors (PRRs), tumor necrosis factor receptor (TNFR), or interferon receptor (IFNR). This promotes the transcription and synthesis of NLRP3, pro-caspase 1, pro-IL1β, and pro-IL18. The second signal or the activation signal is initiated by DAMPs and PAMPs that induce the assembly and activation of the inflammasome. DAMPs include (a) lysosomal and endosomal injury, (b) aberrant ionic influxes, (c) mitochondrial injury, and (d) protein aggregates. (e) PAMPs, such as viral proteins and

RNA, are detected by DAI/ZBP1, DHX33, OAS, or DDX19A leading to the activation of NLRP3 inflammasome. This results in the autocleavage of pro-caspase 1 and the resulting caspase 1 mediates the cleavage of pro-IL1β, pro-IL18, and gasdermin D (GSDMD) into their active forms. (Source: Chunyuan Zhao & Wei Zhao, 2020)

Figure 2.23. (A) retinoic acid-inducible gene I (RIG-I) and (B) melanoma differentiation-associated gene 5 (MDA5) both have two CARD domains at their N-terminal region, two centrally located helicase domains (Hel1 and Hel2), and a carboxy-terminal domain (CTD). The CTD (purple) binds the viral dsRNA (light green) in both (C) RIG-I and (D) MDA5. (E) The RLR CARD domains (blue and aquamarine) interact with the CARD-like domain (orange) of the adaptor molecule mitochondrial anti-viral signaling (MAVS) protein. (Source: Morgan Brisse & Hinh Ly, Creative Commons License)

Figure 2.24. Retinoic acid-induced gene-I-like receptors (RLRs) like RIG-I and melanoma differentiation-associated protein 5 (MDA5) are activated upon binding their viral RNA ligand. They interact with and activate mitochondria antiviral signaling protein (MAVS). MAVS recruits signaling proteins, such as tumor necrosis factor receptor-associated factor 3 (TRAF3), TRAF6, tumor necrosis factor receptor type 1-associated death domain (TRADD), receptor interacting serine/threonine protein kinase 1 (RIP1), and Fas-associated protein with death domain (FADD). TRAF3 activates TANK binding kinase 1 (TBK1) and IκB kinase ε (IKKε), which phosphorylate interferon regulatory factor 3 (IRF3) and IRF7 causing them to dimerize. The resulting dimer goes into the nucleus to induce type I interferon (IFN) response. TRAF6 ubiquitinates NF-κB essential modulator (NEMO), which activates IκB kinase. This, in turn, activates NF-κB, which drives the expression of type I IFN and pro-inflammatory cytokines. (Source: Hui Yee Yong & Dahai Luo, Creative Commons License)

Figure 2.25. Transmembrane C-type lectin receptors (CLRs) contain at least one C-type lectin domain (CTLD), a stalk region, transmembrane domain, and an intracellular or cytoplasmic domain. Type I transmembrane CLRs have their N-terminus at the extracellular side, such as mannose receptor and DCL-1. Meanwhile, the N-terminus of type II is in the cytoplasm like in Dectin-1 and DC-SIGN. The cytoplasmic domain of these two has a tyrosine (Y)-based signaling motif. The monomer of soluble CLRs like mannose-binding lectin (MBL) and surfactant protein-D (SPD) has CTLDs and α helical coils. Two MBL trimers can oligomerize to form a bouquet-like complex, while SP-D forms a cruciform dodecamer. (Source: Ann M. Kerrigan & Gordon D. Brown, Creative Commons License)

Figure 2.26. C-type lectin receptors (CLRs) can have either an activating or inhibiting effect depending on the signaling motif they have in their cytoplasmic domain. When the immunoreceptor tyrosine-based activation motif (ITAM) is activated, the SYK kinase family is recruited and activated. Then PKσ activates the CARD9-Bcl10-Malt1 complex that triggers the NF-κB pathway and subsequent expression of cytokines and chemokines. The ITAM of Dectin-1 may induce an alternative pathway that is mediated by RAF-1 instead of SYK. SYK also induces reactive oxygen species (ROS) generation and inflammasome activation resulting in interleukin-1β (IL-1β) production. Activation of immunoreceptor tyrosine-based inhibition motif (ITIM) triggers the recruitment and activation of protein tyrosine phosphatases like SHP1 or SHP2. These dephosphorylate the motifs resulting in the inhibition of signaling pathways that are mediated by other immune receptors. Other CLRs like CLEC1 have unknown signaling motifs. (Source: Elise Chiffoleau, Creative Commons License)

Figure 2.27. C-type lectin receptors (CLRs) recognize damage-associated molecular patterns (DAMPs). CLEC8A binds oxidized low density lipoproteins (oxLDL) leading to reactive oxygen species (ROS) generation. This triggers IL-1β production. Sin3A associated protein 130 (SAP130) from necrotic cells is recognized by MINCLE resulting in the activation of the immunoreceptor tyrosine-based activation motif (ITAM). This leads to the recruitment and activation of the SYK kinase family and the subsequent activation of the CARD9-Bcl10-Malt1 complex, which induces NF-κB signaling activation, or PLCγ2 that activates the NFAT signaling pathway. Both pathways induce the transcription of cytokine and chemokine genes. CLEC12A interacts with monosodium urate (MSU) crystals, which activates the immunoreceptor tyrosine-based inhibition motif (ITIM). This triggers the recruitment and activation of protein tyrosine phosphatases, SHP1 and SHP2, that dephosphorylate activation motifs. This inhibits cellular activation mediated by other pattern recognition receptors (PRRs). CLEC9A detects F-actin on cell debris. Its hemi-ITAM (HITAM) is involved in facilitating $CD8^+$ T cell cross-priming. CLEC9A can also activate SHP1 to restrain excessive immune response. (Source: Marion Drouin *et al.*, Creative Commons License)

Figure 2.28. The innate immune system comprises the anatomic or surface defenses, physiologic defenses as well as internal defenses made up of phagocytic cells and cells involved in the inflammatory response. It complements adaptive immunity in order to enhance the over-all immune response. (Source: Lindsay M. Biga *et al.*, Creative Commons License)

Figure 2.29. The skin's capacity to protect the body from stressors in the external environment is due to its complex structure and its inherent immune system. The outermost epidermis is made up mostly of keratinocytes (KCs), which are arranged in several sublayers or strata, namely the most external stratum corneum, followed by stratum granulosum, stratum spinosum, and stratum basale. Langerhans cells (LCs) are found in the inner strata of the epidermis, while the melanin-producing melanocytes (blue-colored cells) occur in the innermost stratum basale. The dermis is home to blood and lymphatic vessels as well as a variety of immune cells, such as plasmocytoid dendritic cells (pDCs), dermal dendritic cells (dDCs), macrophages (MØs), natural killer cells (NKs), innate lymphoid cells (ILCs), and T cells. (Florence Abdallah *et al.*, Creative Commons License)

Figure 2.30. Pathogenic microorganisms can penetrate the epidermis when the skin is injured. The pathogen-associated molecular patterns (PAMPs) in these microbes (e.g., *Staphylococcus aureus*) and the damage-associated molecular patterns (DAMPs) from damaged cells are recognized by Toll-like receptors (TLRs), such as TLR2/TLR6. This induces the dendritic cells to express pro-inflammatory cytokines like IL-1β, TNFα, and IFNγ. These recruit neutrophils and macrophages to the site of injury and promote AMP production, such as LL37 and hBD2. TLR3 is specifically activated by viral

PAMPs, resulting in IL27 production. IL27, in turn, induces the expression of antiviral interferon-stimulated genes (ISGs), such as oligoadenylate synthetase 2 (OAS2), by keratinocytes. Pathogenic bacteria inhibit AMP production, while commensal bacteria like *Staphylococcus epidermidis*, promote it. Viruses may also inhibit ISG production. (Source: Margaret Coates *et al.*, Creative Commons License)

Figure 2.31. The skin immune system interacts with skin microbiota to protect against pathogens and to tolerate commensal microorganisms. In a healthy state, *Staphylococcus epidermidis* can control *Staphylococcus* pathogenicity. It stimulates T cells to produce IL17 and IFNγ, therefore inhibiting *Leishmania major* proliferation. *S. aureus* produces σ toxins that trigger local allergic responses in the skin activating Th2-mediated inflammatory response. (Source: Florence Abdallah *et al.*, 2017)

Figure 2.32. A mucosal layer and a single layer of intestinal epithelial cells (IECs) separate the gut microbiota from the underlying tissues. There are intraepithelial lymphocytes (IELs) between the epithelial cells that maintain the mucosal barrier and provide protection from pathogens. Internal to the epithelium is the lamina propria, which is a connective tissue composed of stromal cells, blood vessels, and nerves. It is also made up of immune cells, namely macrophages (blue myeloid cells) and dendritic cells (green myeloid cells). Other innate immune cells in this tissue layer include mast cells, monocytes, neutrophils, eosinophils, T cells (i.e., subsets of CD4+ T cells: Treg, Tr1, Th1, Th2, and Th17), IgA-producing B cells, and innate lymphoid cells (i.e., ILC1, ILC2, and ILC3). (Source: Sara M. Parigi *et al.*, Creative Commons License)

Figure 2.33. The gut-associated lymphoid tissues (GALT) include Peyer's patches and mesenteric lymph nodes (MLN). Microfold (M) cells in the epithelial layer covering GALT sample antigens, which are then ingested by dendritic cells (DCs) and subsequently trigger specific T and B cell responses in the Peyer's patches and MLNs. Retinoic acid produced by DCs enhances the expression of homing receptors (e.g., $\alpha_4\beta_7$ and CCR9) on lymphocytes, which are then guided to enter major effector sites like the lamina propria. (Source: Jerry R. McGhee & Kohtaro Fujihashi, Creative Commons License)

Figure 2.34. When bacterial invasion or epithelial injury is detected, such as in the case of inflammatory bowel disease (IBD), epithelial cells increase the production of α-defensins and β-defensins. The number of macrophages also dramatically increases, and these macrophages express costimulatory molecules of the inflammatory process. The Toll-like receptors (TLRs) on dendritic cells (DCs) are also activated so that there is increased expression of pro-inflammatory cytokines. DCs are also involved in the activation of naïve T cells into the pro-inflammatory Th1, Th2, and Th17 cells as well as the anti-inflammatory Treg cells. Moreover, they are involved in the maturation of B cells in order to produce antibodies such as IgA. (Source: Francesa A.R. Silva *et al.*, Creative Commons License)

Figure 2.35. The respiratory tract is protected by physical barriers like the epithelial layer made up of ciliated cells (CC) that are firmly connected by junctional complexes, one component of which is the tight junction (TJ). These complexes seal the paracellular spaces so that passage of harmful pathogens and substances is prevented. Another barrier is formed by mucins (green), secreted by goblet cells (GC), and glycolipids that form a layer of glycocalyx over the cilia, which can protrude up to 1,500 nm from the apical surface of the epithelium. This restricts the access of substances from the lumen to the apical cell surface so that large-sized pathogens can be cut off from their receptor (inset). Mucins with absorbed water also form a viscous gel or mucus that float on the surface barriers and trap microbes and particulate matters so that these are transported upward through the airway lumen via the coordinated beating of the cilia. (Source: Andreas Frey *et al.*, Creative Commons License)

Figure 2.36. The respiratory epithelium made up of bronchial and alveolar epithelial cells (BECs and AECs) provides a protective barrier against pathogens and particulates in inhaled air. It is complemented by immune cells present in the respiratory system, namely macrophages, dendritic cells (DCs), subsets of innate lymphoid cells (ILC1, ILC2, and ILC3), and neutrophils. Lungs also have bronchus-associated lymphoid tissue (BALT), which contains B cells, T cells, and DCs, and is induced to become iBALT in response to infection. (Source: Vijay Kumar, Creative Commons License)

Figure 2.37. There are three groups of human Fcγ receptors, namely FcγRI, FcγRII, and FcγRIII. FcγRI has an α subunit with three Ig-like extracellular domains. Meanwhile, the α subunit of the other two types has two of these Ig-like domains. FcγRI and FcγRIIIa have an accessory dimer of γ chains which contain an immunoreceptor tyrosine-based activation motif (ITAM). FcγRIIa has an ITAM in its α subunit. FcγRIIb, meanwhile, has an immunoreceptor tyrosine-based inhibition motif (ITIM) in its α subunit. Lastly, FcγRIIIb is bound to the plasma membrane via a glycosylphosphatidylinositol (GPI) anchor. (Source: Eileen Uribe-Querol & Carlos Rosales, Creative Commons License)

Figure 2.38. The three groups of complement receptors are CR1 and CR2, CR3 and CR4, and CRIg. (Source: Eileen Uribe-Querol & Carlos Rosales, Creative Commons License)

Figure 2.39. Most phagocytic receptors like FcγRIIa and integrin CR3 cooperate to bind the antigen to be engulfed. FcγRIIa receptors engage with several IgG antibodies on the target so that these receptors aggregate triggering an inside-out signal that activates CR3 via the GTPase Rap. The activated CR3 assumes an extended conformation allowing it to bind the target via the complement fragment C3b and to form a diffusion barrier that excludes larger transmembrane glycoproteins, such as CD45. (Source: Eileen Uribe-Querol & Carlos Rosales, Creative Commons License)

Figure 2.40. When FcγRIIa engages IgG antibodies on the target particle (antigen), it activates Src family kinase (SFK), which phosphorylates tyrosine residues in the ITAM (green domain) of the receptor. Then spleen tyrosine kinase (Syk) associates with the phosphorylated ITAMs and phosphorylates a signaling complex composed of the scaffold protein linker for activation of T cells (LAT) and other proteins, such as phospholipase C gamma (PLCγ). The latter produces inositoltriphosphate (IP_3) that induces calcium release from the endoplasmic reticulum (ER) and diacylglycerol (DAG) that activates protein kinase C (PKC). PKC activates extracellular signal-regulated kinases ERK and p38. Syk also recruits and activates phosphatidylinositol 3-kinase (PI_3K), which produces phosphatidylinositol-3,4,5-triphosphate (PIP_3) that regulates contractile proteins like myosin. PIP_3 and Vav activate the GTPase Rac, which is involved in the activation of the transcription factors, such as NF-κB and JNK. (Source: Eileen Uribe-Querol & Carlos Rosales, Creative Commons License)

Figure 2.41. When the complement receptor 3 (CR3) binds the complement molecules iC3b on the target particle (antigen), a signaling pathway is initiated, which activates the GTPase Rho. Rho may activate Rho kinase (ROCK), which phosphorylates and activates myosin II leading to the accumulation of Arp2/3 complex that promotes actin polymerization at the phagocytic cup. Rho may also promote the accumulation of mammalian diaphanous-related formin 1 (mDia1) which drives actin polymerization and binds to CLIP-170 providing a link to microtubules. (Source: Eileen Uribe-Querol & Carlos Rosales, Creative Commons License)

Figure 2.42. The phagosome undergoes a three-stage maturation process: early (A), late (B), and the phagolysosome formation (C). As the phagosome fuses with an endosome, its size remains the same because recycling vesicles also bud off from it. However, its membrane composition changes to include molecules that regulate membrane fusion, including Rab5, early endosome antigen 1 (EEA1), and Rab7. The interior of the phagosome also acidifies with the phagolysosome being the most acidic due to the action of the proton-pumping V-ATPase. The phagolysosome also acquires several degradative enzymes (scissors) and, the NADPH oxidase complex in its membrane initiates a pathway that produces the microbicide hypochlorous acid (Uribe-Querol & Rosales, 2020).

Figure 2.43. Inflammation is induced by the detection of foreign entities or tissue injury. Histamine is released causing increased blood flow that leads to the inflamed site becoming red and warm. The influx of immune cells leads to the characteristic swelling. (Source: Lindsay M. Biga *et al.*, Creative Commons License)

Chapter 3

Opening image. Killer cells surround a cancer cell. (Source: National Institutes of Health, Public Domain)

Figure 3.1. A T cell that expresses CD4 can be activated by antigen presenting cells (APCs) that present the ingested antigen via a major histocompatibility complex (MHC) molecule to the T cell receptor (TCR). This leads to the conversion of the T cell into a helper T cell. (Source: Sjef, Creative Commons License)

Figure 3.2. A T cell expressing CD8 may be activated by antigens presented by major histocompatibility complex class I (MHC I) on an antigen presenting cell (APC) so that it matures into a cytotoxic T cell. (Sjef, Creative Commons License)

Figure 3.3. Immature B cell leaves the bone marrow and migrates to lymphoid organs, such as the spleen, where it matures into a transitional B cell (T1 and T2) that further differentiates into either a marginal zone B cell or a mature (follicular) B cell. (Source: Bobologist, Creative Commons License)

Figure 3.4. A naïve B cell may be stimulated by antigenic exposure and cytokines to develop into either short-lived plasma cell (plasmablast), long-lived plasma cell or memory cell. (Source: Bobologist, Creative Commons License)

Figure 3.5. (A) An immunoglobulin is made up of two light chains and two heavy chains having a Y-like conformation. The heavy chains are linked by a disulfide bond (yellow) between two cysteine residues (C-C) at the hinge region (black). Each light chain is also joined to the heavy chain via disulfide linkages. Each chain is composed of one variable (V) as well as one constant domain in the light chain (CL) and two to three constant domains in the heavy chain (CH). When an immunoglobulin is cleaved by a protease, it produces the crystallizable fragment (Fc) and two antigen binding fragments (Fab), which, in turn, can be further split into the variable (Fv) and constant regions. (B) Fv is made up of VH and VL domains, each of which has three complementarity determining regions (CDRs). (Source: Aleksandr Kovaltsuk *et al.*, Creative Commons License)

Figure 3.6. (Upper) The monomeric and membrane-bound IgM contains Fab (antigen-binding) and Fc (crystallizable) fragments as well as a transmembrane signaling tail that binds to Fc receptors on the surface of B cells. (Lower left) The pentameric IgM has 10 antigen binding sites in its variable regions (green), and it is characterized by the presence of a joining chain (J-chain). This chain is absent in the (lower right) hexameric form. (Source: Upper image- Katelyn Jones *et* al., Creative Commons License; Lower images-Bruce A. Keyt *et al.*, Creative Commons License)

Figure 3.7. Both IgM and IgG can activate the complement system leading to complement-dependent cytotoxicity (CDC). IgM's hexameric or pentameric structure provides multiple antigen binding sites and enables highly avid binding to the complement component C1q. Thus, IgM can fix complement significantly better than IgG. (Source: Bruce A. Keyt *et al.*, Creative Commons License)

Figure 3.8. IgM can bind at least three types of receptors, namely the polymeric Ig receptor (pIgR), Fcα/μ receptor (Fcα/μR), and FcμR. They differ in terms of their tissue distribution, their requirement for binding IgM, and the number of their immunoglobulin fold-like regions (oval). (Source: Bruce A. Keyt *et al.*, Creative Commons License)

Figure 3.9. Immunoglobulin has four subclasses, namely IgG1, IgG2, IgG3, and IgG4. These differ in certain amino acid residues (small blue stars) in the constant domain of the heavy chains, in the hinge region (large blue star) as well as in their glycosylation. (Source: Steven W. de Taeye *et al.*, Creative Commons License)

Figure 3.10. (A) Immunoglobulin A (IgA) has two subtypes, IgA1 and IgA2, made up of two heavy chains (blue) and two light chains (brown). IgA1 has a longer hinge with O-linked glycosylation, while IgA2 has more N-linked glycosylation sites. (B) Each subtype may assume monomeric and dimeric conformations. The dimeric form consists of two monomers joined together by a joining chain (J-chain) (green). Secretory IgA is mainly dimeric, and it is associated with the secretory component (red). (Source: Annelot Breedveld & Marjolein van Egmond, Creative Commons License)

Figure 3.11. IgA is transported across the mucosal epithelium after binding to the polymeric Ig receptor (pIgR). (1) Dimeric IgA (red) produced at the lamina propria of the epithelium binds pIgR (cyan) at the basolateral surface of the epithelial cell. (2) The pIGR-IgA complex is internalized and transported via vesicles across the cell. (3) The secretory component (SC) of the pIgR is cleaved and then associates via a disulfide bond to the dimeric IgA. (4) At the apical surface, the IgA is secreted (i.e., secretory IgA, sIgA) into the lumen. (5) It then binds and neutralizes the pathogens (purple and dark blue). (6) Some pathogens (pink) may penetrate the epithelium and reach the lamina propria. (7) These are bound by dimeric IgA. (8) The pathogen-IgA complex binds to pIgR. (9) After phagocytosis and vesicular transport across the cell, the pathogen is released back into the lumen. (10) Some intracellular pathogens (lime green) may be intercepted by dimeric IgA being transported across the cell. (11) The pathogen is ejected upon release of sIgA at the lumen. (12) Dimeric IgA can also engage phagocytes to engulf and clear pathogens. (Source: Source: Patricia de Sousa-Pereira & Jenny M. Woof, Creative Commons License)

Figure 3.12. The structure of (a) IgG shows some differences compared to that of (b) IgE. The latter has two additional constant domains in each of its heavy chains and it lacks a hinge region. (Source: Brian J. Sutton *et al.*, Creative Commons License)

Figure 3.13. An allergen (green) can engage two identical IgE antibodies (yellow and blue) that are linked to their respective FcεRI (purple). This can trigger a cascade of events that result in hypersensitivity and allergic reactions. (Source: Brian J. Sutton *et al.*, Creative Commons License)

Figure 3.14. (A) The recombination signal sequence (RSS) flanks the gene segments in the locus for the variable region of the immunoglobulin heavy chain (i.e., variable, diversitys, and joining genes) as well as the V and J genes in the λ and κ light chain locus. The RSS in these genes varies in terms of orientation as well as the length of their spacer sequence. (B) The RSS has a conserved heptamer sequence and a conserved nonamer sequence with a non-conserved spacer sequence between them. The space can have either 23 base pairs (bp) or 12 bp. (C) During somatic recombination in the variable locus of the λ light chain locus, RAG1/RAG2 complex binds the 23-bp linker RSS adjacent to the V gene and the 12-bp linker RSS adjacent to the J gene. The complex then cleaves the two RSS along with the sequence between V and J (loop excision) resulting in a VJ joined segment. (Source: Oliver Backhaus, Creative Commons License)

Figure 3.15. The germline IGH (immunoglobulin heavy chain) locus contains several gene segments for the constant (C) and variable (V) regions. Somatic recombination by RAG proteins joins the diversity (J) and joining (J) segments first then the variable (V) segment is joined next. After transcription, C segments in the primary transcript are spliced and one of them is recombined with the VDJ segment to form the final mRNA. (Source: Oliver Backhaus, Creative Commons License)

Figure 3.16. After RAG1/RAG2 removes the intervening sequence between the variable (V) and joining (J) genes of the λ light chain, hairpins are formed at the excision site. The Artemis complex then makes a single-stranded cut in both hairpins producing palindromic DNA sequences. These so-called P nucleotides are extended by the addition of random nucleotides, called N nucleotides, by the terminal deoxynucleotidyl-transferase (TdT). The complementary bases in the two single-strand segments pair together, while the mismatched nucleotides are removed by an exonuclease. The gaps are then filled by a DNA polymerase and the breaks in the DNA strands are ligated.

(Source: Oliver Backhaus, Creative Commons License)

Figure 3.17. B cells in the dark zone of the lymph node germinal center highly express the chemokine receptor CXCR4 and undergo somatic hypermutation in the variable region genes of their immunoglobulin loci. When they enter the light zone, CXCR4 expression is downregulated. Follicular dendritic cells (FDCs) in the light zone present foreign antigens, which are bound by specific B cells with B-cell receptors (BCRs) that have an affinity to them. The resulting antigen/BCR complex is processed, and the generated antigen peptides are presented by major histocompatibility complex II (MHC II) to the T cell receptor (TCR) of follicular helper T (T_{FH}) cells. The CD40-CD40L interaction and cytokines from T cells provide survival and proliferation signals to the B cells. On the other hand, the low-affinity B cells undergo apoptosis. The surviving B cells then reenter the dark zone where they can undergo another cycle of somatic hypermutation. This cyclic affinity maturation process produces B cells with high-affinity BCRs. These B cells finally leave the light zone and differentiate into plasma cells or memory cells. (Source: Oliver Backhaus, Creative Commons License)

Figure 3.18. Class-switch recombination is induced by activation-induced cytidine deaminase (AID) producing a single-stranded break on two switch (S) regions upstream of constant gene segments of the heavy chain locus. The DNA between the cuts is removed and non-homologous end joining machinery (NHEJ) joins the remaining constant gene segments to the V(D)J segment. (Source: Oliver Backhaus, Creative Commons License)

Figure 3.19. The immunoglobulin paratope (light green) binds to the antigen epitope (orange). (Source: Richard A. Norman *et al.*, Creative Commons License)

Figure 3.20. A proteinaceous antigen is typically composed of multiple epitopes so that it can be recognized by different immunoglobulins or antibodies. (Source: OpenStax, Creative Commons License)

Figure 3.21. (A) The Y-shaped arm of the immunoglobulin is known as the antigen binding fragment (Fab), which is composed of the two variable domains of the heavy (VH) and light chains (VL) as well as the two constant domains (CH1 and CL). VH and VL form the Fv fragment. (B) This fragment contains the paratope consisting of the three hypervariable loops in the VL (i.e., L1, L2, and L3) and three in the VH (i.e., H1, H2, and H3). (Source: Inbal Sela-Culang *et al.*, Creative Commons License)

Figure 3.22. The paratope of an immunoglobulin may bind the entire antigen via the conformational epitope. Alternatively, a linear ectopic element of the said epitope may be bound in a similar manner but with fewer points of interaction (stars). (Source: Line Ledsgaard *et al.*, Creative Commons License)

Figure 3.23. The T cell-dependent activation of B cells is initiated by a first signal generated by the cross-linking of B cell receptors (BCRs) upon antigen binding. The antigen is then ingested, processed, and presented by the B cell as peptide-major histocompatibility class II (MHC II) complexes, which are recognized by the T cell receptor (TCR) and CD4 of the helper T (Th) cell. The subsequent interaction of the B cell CD40 and the CD40L of the Th cell provides the second signal. Thereafter, the

Th cell produces cytokines (e.g., IL2, IL4, IL5) to activate the B cells to proliferate. (Source: Altaileopard, Public Domain)

Figure 3.24. The T-dependent activation of B cells leads to the multiplication of B cells and their differentiation to plasma cells that secrete immunoglobulins or to memory cells. (Source: Charles Molnar & Jane Gair, Creative Commons License)

Figure 3.25. There are various mechanisms initiated by immunoglobulins to protect the host from pathogens. They can initiate the activation of the complement system resulting in (1) bacteriolysis or bursting of the bacterium or (2) opsonization that triggers phagocytes to ingest the pathogen. They also mediate the destruction of pathogenic cells by natural killer (NK) cells through a process called (3) antibody-dependent cell-mediated cytotoxicity (ADCC). (4) Agglutination or the aggregation of antigen-antibody complexes is essential in initiating opsonization and ADCC. Other mechanisms include the (5) neutralization of the pathogen or its toxin by the binding of the immunoglobulins and (6) secretion blockade that prevents type III secretion system (T3SS) proteins from being secreted. (Source: Julia A. Hotinger & Aaron E. May, Creative Commons License)

Figure 3.26. Antibody-dependent cell-mediated cytotoxicity (ADCC) is one mechanism by which helminth parasites are destroyed. Immunoglobulins interact with the parasite and their Fc is bound by Fc receptors found on macrophages, neutrophils, and eosinophils. This triggers the degranulation of these cells and the subsequent release of substances that lyse the parasite, such as nitric oxide (NO), hydrogen peroxide (H2O2), and toxic proteins like major basic protein (MBP), eosinophil peroxidase (EPO), eosinophil cationic protein (ECP), and eosinophil-derived neurotoxin (EDN). (Source: Claudia Cristina Motran *et al.*, Creative Commons License)

Figure 3.27. Maternal IgG is taken up into the syncytiotrophoblast cell of the placenta via endocytosis. The neonatal Fc receptor (FcRn) at the inner membrane of the acidic endosome binds two IgG molecules. Upon fusion of the endosome with the basolateral membrane, FcRn releases the IgGs due to the increased pH of the extracellular fluid at the fetal side of the placenta. The FcRn can then be recycled to perform another transport cycle. (Source: Marie Albrecht & Petra Clara Arck, Creative Commons License)

Figure 3.28. The polymeric Ig receptor (pIgR) binds the joining chain of the dimeric IgA molecule at the basolateral side of the mammary epithelial cell. The pIgR-IgA complex is then taken in by endocytosis. When the endosome fuses with the apical membrane of the cell, the pIgR releases the secretory form of the IgA (sIgA) into the lumen that contains the breastmilk. (Source: Marie Albrecht & Petra Clara Arck, Creative Commons License)

Figure 3.29. Among the four types of hypersensitivity reactions, the first three involve immunoglobulins while type IV is T cell-mediated. Type I is mediated by IgE that leads to degranulation of mast cells and basophils so that they release molecules that stimulate the allergic response. Type II involves IgG and ,in some cases, IgM that bind to self-antigens activating antibody-dependent cellular cytotoxicity or the complement system. Type III is characterized by the formation of immune complexes consisting of antigens bound by immunoglobulins that eventually activate the complement system and neutrophil influx. Type IV response is initiated by a sensitized helper T cell that produces cytokines resulting in activation of macrophages or cytotoxic T cells (Source: J. Gordon Betts *et al.*, Creative Commons License)

Figure 3.30. (a) A T cell receptor (TCR) interacts with an antigen bound by a major histocompatibility complex (MHC) molecule on the surface of an antigen presenting cell (APC). (b) The TCR α chain (TRA) locus on chromosome 14 has an array of several variable (V) and joining (J) gene segments as well as a constant (C) gene segment. The TCR β chain (TRB) locus on chromosome 7 has the same kinds of gene segments with an additional diversity (D) segments. During T cell development, the V(D)J segments in the germline DNA are rearranged to form the final rearranged DNA sequence. Following transcription, the sequence between the recombined V(D)J regions and the C region is spliced out. The complementarity-determining region (CDR) 1 and CDR2 are encoded within the germline V gene segment, while the CDR3 is coded for by the V(D) J junction. (Source: Elisa Rosati *et al.*, Creative Commons License)

Figure 3.31. The T cell receptor (TCR) is a heterodimer (αβ) located on the T cell membrane with a variable (V) and a constant (C) domain in each chain. It is associated with a complex of CD3 proteins forming three dimers, namely epsilon-delta (εδ), epsilon-gamma (εγ), and zeta-zeta (ζζ). (Source: Pappanaicken R. Kumaresan *et al.*, Creative Commons License)

Figure 3.32. Two signals are required in order to fully activate T cells. Signal 1 is provided by the interaction of TCR and the peptide-major histocompatibility complex (MHC). The binding of CD4 or CD8 with the MHC stabilizes such interaction. Signal 2 is provided by the binding of co-stimulatory receptors (e.g., CD28, CD40L, LFA-1) with their corresponding ligands. An inhibitory signal may also be transmitted via the interaction of CTLA-4 with B7. (Source: Yu Tai *et al.*, Creative Commons License)

Figure 3.33. The CD4 or CD8 serve as co-receptors on the T cell membrane. They have extracellular domains that engage the major histocompatibility complex (MHC) to stabilize the peptide-MHC interaction. Their cytoplasmic tails associate with the kinase Lck. Also, there are co-stimulatory receptors on the T cell surface, such as CD28 and 4-1BB. (Nicholas J. Chandler *et al.*, Creative Commons License)

Figure 3.34. Naïve CD4 T cells are activated by two signals. The first one is produced when the T cell receptor (TCR) binds a peptide presented by the major histocompatibility complex II (MHC II) molecule on the surface of an antigen presenting cell (APC). The CD4 co-receptor stabilizes the TCR-MHC II interaction. The second signal is provided by the interaction of co-stimulatory molecules on the APC like B7.1 and B7.2 with CD28 on the T cell. Finally, cytokines from macrophages and dendritic cells stimulate the CD4 T cell to differentiate into helper T (Th) cells. (Source: Hope O'Donnell & Stephen J. McSorley, Creative Commons License)

Figure 3.35. Naïve CD4 T cells differentiate into various types of helper T (Th of Tfh) cells or into regulatory T (Treg) cells through the action of different combinations of cytokines. In turn, each type of activated CD4 T cells produces a distinct set of cytokines. (Source: Vita Golubovskaya & Lijun Wu, Creative Commons License)

Figure 3.36. The first signal that activates CD8 T cells is provided by their recognition of peptide-major histocompatibility complex I (pMHC I) complex on an antigen presenting cell like a migratory conventional dendritic cell (see inset). A complementary signal is provided by co-stimulatory molecules like 4-1BB, CD28, and CD27. These two signals along with certain cytokines stimulate the CD8 T cells to proliferate and enhance their cytotoxic functions. After proliferation, many of the T cells undergo apoptosis leaving a few that differentiates into different types of memory cells. (Source: Marjorie Schluck *et al.*, Creative commons license; inset: Julie Busselaar *et al.*, Creative Commons License)

Figure 3.37. In lymph tissues, CD4 T cells are initially activated upon recognition of peptides presented by antigen presenting cells (APCs) and are further activated by a second signal. During intracellular infections, the cytokines IFNγ and IL12 trigger the differentiation of the activated CD4 T cells into helper T cells 1 (Th1), which produce IFNγ. Then the Th1 cells move to the infection site where they are further stimulated to produce effector cytokines, which ultimately results in the destruction of the infected cell. (Source: Source: Hope O'Donnell & Stephen J. McSorley, Creative Commons License)

Figure 3.38. CD8 T cells can induce cytolysis by two mechanisms. One is by releasing granules containing perforin that creates pores in the target cell membrane and granzyme B that induces DNA and protein fragmentation. Another way is via the Fas-FasL interaction that initiates a cascade that activates caspase enzymes, which mediate apoptosis. CD8 T cells also secrete cytokines, such as IFNγ, TFNα, and IL17, and the chemokine CCL5 that regulate antiviral and inflammatory processes. (Source: Federico Perdomo-Celis *et al.*, Creative Commons License)

Figure 3.39. (a) In the secondary lymphoid organs, antigen activation leads to B- and T-cell interaction at the boundary of the B- and T-cell zones. (b) The activated B cells proliferate and differentiate either into (1) plasma cells, (2) very early memory B cells, or (3) cells with upregulated BCL6 that form a germinal center (GC). (c) In the light zone of the GC, high affinity B cells receive strong T-cell support resulting in a transcriptional profile that produces plasma cell precursors. (d) These precursors either enter circulation as short-lived plasma cells or migrate to the bone marrow, where they develop into long-lived plasma cells. (e) The low affinity B cells in the light zone receive weaker T-cell help. Thus, they will follow a different transcriptional program, which leads to memory B cell development. (f) These will then reside in secondary lymphoid organs, become tissue residents, or enter circulation. (Source: Anna-Karin E. Palm & Carole Henry, Creative Commons License)

Figure 3.40. (a) The antibodies produced by long-lived plasma cells (purple) provide the first line of defense against secondary antigen exposure. (b) If the antigen persists, then tissue-resident and circulating memory B cells (green) are directly activated. Some of the antigens are carried to secondary lymphoid organs. The antigen activates memory B cells in lymph nodes to form subcapsular proliferative foci (SPF), a germinal center (GC), or plasma cells (purple). (c) SPF produce plasma cells, new memory B cells, and B cells that enter the germinal center (GC). (d) GCs are usually formed by unmutated, low affinity, IgM$^+$, CD80$^-$ PD-L2$^-$memory B cells, but some of these may become plasma cells. (e) Most of the IgM$^+$ and switched memory B cells that express either CD80 or PD-L2 directly develop into plasma cells, although they may also form GCs. (f) Switched, high affinity memory B cells that express both markers exclusively form new plasma cells. (Source: Anna-Karin E. Palm & Carole Henry, Creative Commons License)

Figure 3.41. There are various types of memory T cells. Stem cell memory (Tscm) cells and central memory T (Tcm) cells are found in lymphoid organs and in the blood. They are relatively undifferentiated and long-lived. Effector memory T (Tem) cells and tissue-resident memory T (Trm) cells both occur in peripheral tissues, but Tem cells migrate between tissues and the blood, while Trm cells are restricted in tissues. They are more differentiated than Tscm and Tcm cells. Follicular helper T (Tfh) cells are effector CD4 T cells positioned near B cell follicles to assist in germinal center formation. (Source: Shafqat Ahrar Jaigirdar & Megan K.L. MacLeod, Creative Commons License)

Figure 3.42. There are different models for memory T cell development. (a) The linear model proposes that when naïve T cells are activated by interacting with antigen-major histocompatibility complex (MHC) conjugate on antigen-presenting cells (APCs), they develop into effector T cells, some of which give rise to effector memory T (T_{EM}) cells and central memory T (T_{CM}) cells. It is uncertain if the precursor memory cells also give rise to tissue-resident memory T (T_{RM}) cells. (b) In the asymmetrical model, the daughter cells proximal to the immune-synapse develop into T_{EM} cells, while those distal daughter cells become T_{CM} cells. The origin of T_{RM} cells is unknown. (c) In the self-renewal model, naïve T cells give rise to self-renewing effector T cells or T_{CM} cells, which give rise to T_{EM} cells. T_{RM} cells may possibly come from the self-renewing cells or from the T_{EM} cells. (d) The simultaneous model proposes that naïve T cells differentiate into different types of T cells, which then differentiate into different types of memory cells. Helper T cell 1 (Th1) and Th17 become T_{EM}cell, while follicular helper T (T_{FH}) cell gives rise to T_{CM} cell. The Th cell type that differentiates into T_{RM} is still unknown. (Source: Itay Raphael *et al.*, Creative Commons License)

Figure 3.43. CD4 effector T cells (T_{EFF}) or regulatory cells ($Treg_{EFF}$) may differentiate into central memory T cells (T_{CM}). T_{EFF} may also differentiate into effector memory T (T_{EM}) cells, whereas Treg cells may develop into Treg effector memory ($Treg_{EM}$) cells. During differentiation, they acquire or lose the expression of certain cell surface markers. (Source: Vita Golubovskaya & Lijun Wu, Creative Commons License)

Figure 3.44. Naïve CD8 T cells may differentiate into memory stem T (T_{SCM}) cells, which in turn, can develop into central memory T (T_{CM}) cells, then into effector memory T ($_{TEM}$) cells. Each type of CD8 memory T cell expresses a distinct set of cell surface markers. Also, the potential to proliferate decreases as the T cell differentiates, whereas its effector function gets more enhanced. (Source: Vita Golubovskaya & Lijun Wu, Creative Commons License)

Figure 3.45. High levels of interleukin 2 (IL2) are vital for the development of naïve CD4 T cells into memory T cells. Such development is dependent on the level of inflammatory cytokines, TCR affinity, and precursor frequencies. Increased levels of inflammatory signals favor the formation of effector memory T (T_{EM}) cells, while increased TCR signaling and precursor frequencies lead to the formation of central memory T (T_{CM}) cells. TGF-β is known to be vital for tissue-resident memory (T_{RM}) cells, but the involvement of other inflammatory signals and TCR signaling remains unclear. (Source: Itay Raphael *et al.*, Creative Commons License)

Figure 3.46. Differential expression of certain regulatory proteins is observed as T cells develop into different types of memory T cells. The expression of EOMES and TCF1 is upregulated for central memory T (T_{CM}) cells. Meanwhile, KLRG1 and Tbet are highly expressed in effector memory T (T_{EM}) cells and KLGR1 expression is observed in tissue-resident memory T (T_{RM}) cells. T_{CM} cells can give rise to T_{EM} cells and vice versa. (Source: Itay Raphael *et al.*, Creative Commons License)

Figure 3.47. (1) Tissue resident memory T (T_{RM}) cells are likely derived from effector T cells that are recruited by certain chemokines to migrate into the target tissues. Most of the effector T cells die but some differentiate into various types of resting memory T cells including central memory T (T_{CM}) cells (migrate back to lymphoid tissues), effector memory T (T_{EM}) cells (circulate through peripheral tissues), and T_{RM} cells. (2) The T_{RM} cells are retained in the mucosal tissues possibly via the inhibition of the sphingosine-1-phosphate receptor 1 (S1PR1) that promotes lymphocyte exit from tissues and via the expression of integrins that facilitate cell-to-cell interactions. (3) The homeostasis of T_{RM} cells may depend on cytokines that maintain their survival, constitutive low-level inflammation, and, in some cases, the persistence of antigen at the site. (Source: Damian Lanz Turner & Donna L. Farber, Creative Commons License)

LIST OF TABLES

PREFACE

Getting sick is a fact of life. Still, we have the ability to survive various diseases and injury. Almost everyone has an awareness of the immune system that is responsible for such ability. Perhaps some have a passing curiosity about the immune response, others may want to know more about the intricacies of this process. In both cases, this book, **Fundamentals of the Immune System**, can prove to be helpful. This book aims to orient the reader about the various parts of the human immune system and how they work together to protect the body from pathogens.

There are various topics that can be covered under immunology. However, this book does not attempt to explain everything about human immunological processes but focuses on the two main types of immune responses —innate and adaptive immunity. A whole chapter is devoted to the discussion of each type providing ample information for the reader to get a deeper understanding of each process. Certain aspects of the immune system are still under intense investigation and so some information presented in this book are possibilities supported by experimental data and not yet established facts.

Figures and tables are provided to complement and supplement the textual explanations. A careful effort is also made to make the language as simple as possible. Abbreviations and acronyms are also employed to streamline the discussion, although most of these are spelled out the first time these are mentioned in the book. A complete list of the references used in the discussion is also provided to allow the reader to do his own further reading and research about certain topics in the book.

With pandemics becoming a part of the human experience nowadays, people in general have heightened interest in the microbial threat they are facing and how this can possibly be defeated. Unfortunately, misleading information abounds regarding how some of these diseases can be treated. This book does not, by any means, promote any medical procedure or therapy, but it does provide well-researched information that can help readers attain a better comprehension of the body's amazing ability to protect and heal itself.

CHAPTER 1

INTRODUCTION TO THE IMMUNE SYSTEM

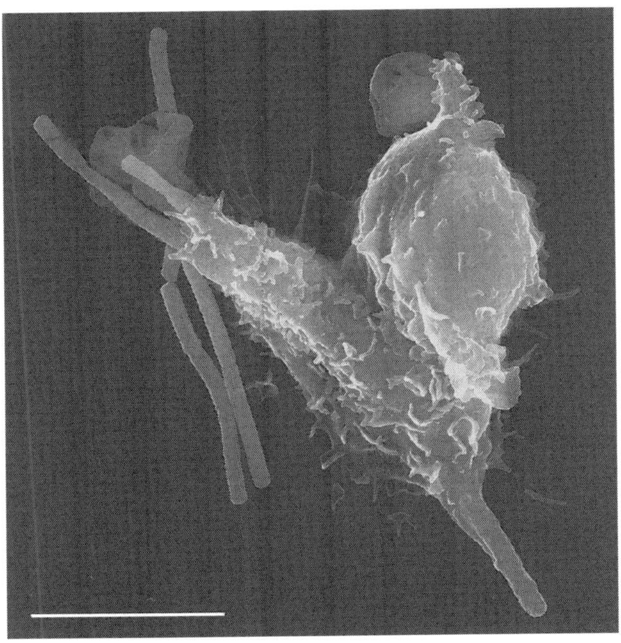

(Source: Volker Brinkmann, Creative Commons License)

Immunity is a well-known concept that is usually associated with the body's ability to fight infections and diseases. It is a function of the immune system which is a complex network of cells and proteins that battle various pathogens. For example, certain cells in the blood called neutrophils can engulf and destroy a whole bacterium that invaded the body (see opening image). A broader definition describes the immune system as a mechanism that enables an organism to distinguish between "self" and "nonself" or to exclude external life from the internal environment. It is present in multicellular organisms that range from the simple sea sponge to the more complex animals and humans (McComb *et al*., 2019).

1.1 SIGNIFICANCE OF THE IMMUNE SYSTEM

The human body faces many disease-causing and dangerous molecules. Some of its own cells may also undergo changes that can lead to illness. The immune system protects the body from these harmful things and processes. Specifically, it fights disease-causing organisms such as viruses, bacteria, fungi or parasites and eliminates them from the body (see Figure 1.1). Also, it detects and neutralizes toxic substances from the environment. The immune system also fights changes in the body that may lead to diseases such as cancer (Institute for Quality and Efficiency in Health Care, 2020).

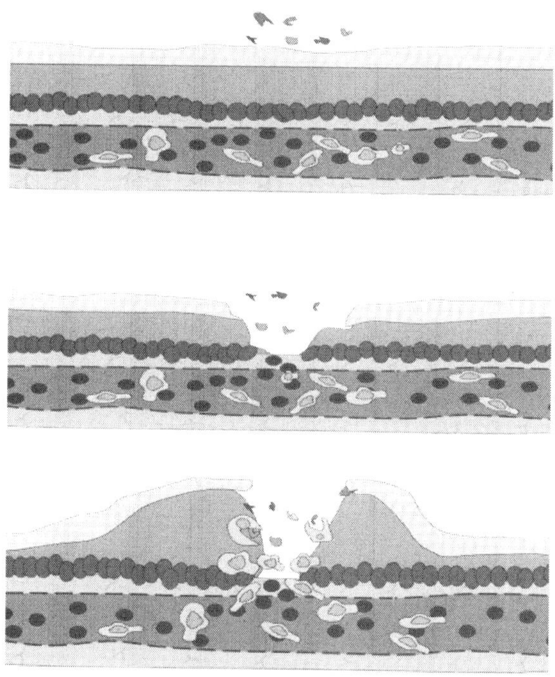

Figure 1.1. A cut on the skin may serve as a gateway for harmful microorganisms to enter the body. Certain white blood cells called phagocytes (light blue) can swallow and destroy these foreign invaders. (Source: Nason Vassiliev, Creative Commons License)

The immune system is activated by antigens, which are various things that the body does not recognize as its own. The proteins on the surfaces of disease-causing microorganisms are examples of antigens. These are recognized and attached to receptors on immune cells triggering a cascade of events in the body to fight the antigen. When the antigen is encountered by the immune system for the first time, it usually stores information about the antigen and how to combat it. During subsequent encounters with the same antigen, the immune system can immediately recognize it and can launch an immune response faster (Institute for Quality and Efficiency in Health Care, 2020).

The body's own cells have surface proteins also, but these do not normally trigger an immune response. However, there are cases when the immune system considers the body's own cells as foreign and then attacks them. This is called an autoimmune response (Institute for Quality and Efficiency in Health Care, 2020). Apart from autoimmune disorders, immune system malfunction can contribute to many diseases like allergies, immunodeficiencies, chronic inflammation, and cancer.

1.2 OVERVIEW OF THE TYPES OF IMMUNE RESPONSES

There are two distinct subsystems of immunity, the innate and adaptive immune systems (see Figure 1.2). While these two branches have different components and mechanisms, they are interdependent or complementary. Innate immunity is the first line of defense against an intruding pathogen that is utilized by the body immediately or within hours of encountering an antigen. On the other hand, adaptive immunity takes a while, taking a week or so, before a maximal response is implemented against the pathogen (Marshall *et al.*, 2018). The succeeding chapters provide more details about these two types of immune systems.

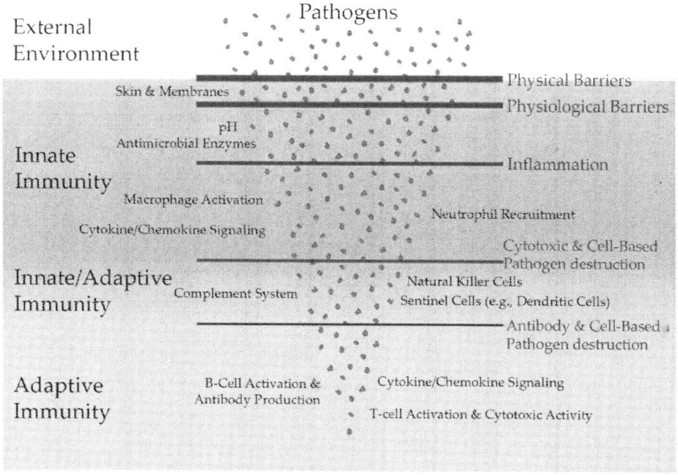

Figure 1.2. The immune system branches into the innate and adaptive systems. The innate immunity includes physical and physiological barriers, cells that detect and attack pathogens, small proteins that signal infection, and short peptides that attach to and restrict microbes. The adaptive immune system consists of cells and proteins that recognize and eliminate specific pathogens and memory immune cells that provide long-term protection. Both innate and adaptive immune systems employ cytokine/chemokine signaling to recruit more immune cells (Source: Martin J. Spiering, Public Domain)

The innate immune system is also called non-specific since it is antigen-independent. It involves protective barriers and immune cells like natural killer (NK) cells and phagocytes (e.g., macrophages and neutrophils) as well as small proteins that signal pathogen invasion (e.g., cytokines and chemokines) (see Figure 1.2). It mainly protects from toxins or pathogens that enter the body through the skin or via ingestion (Institute for Quality and Efficiency in Health Care, 2020; Spiering, 2015).

In the case of the adaptive immune system, it is also known as the specific or acquired immune response since it targets a specific pathogen that invaded the body. Since this is adaptive, it enables the body to fight pathogens, such as bacteria and viruses, that mutate over time (Institute for Quality and Efficiency in Health Care, 2020). It is made up of specialized cells (e.g., B and T cells), proteins (e.g., immunoglobulins), and cells that can store immune memory of a past pathogenic invasion (e.g., memory B and T cells) (see Figure 1.2). The production and storage of these memory cells allow a more rapid response to future invasions by the same pathogen (Spiering, 2015).

Both the innate and the adaptive immune systems utilize cytokine or chemokine signaling to recruit additional immune cells. Cytokines are glycoproteins that serve as intercellular messengers providing soluble regulatory signals that initiate and constrain inflammatory responses to pathogens and injury (Lacy &Stow, 2011; Testar, 2021). On the other hand, chemokines are a family of chemoattractant cytokines that are critical for cell migration from the blood into tissues through the venules and vice versa, and in the induction of cell movement in response to a chemical gradient through a process known as chemotaxis (Oldham, 2021).

The adaptive immune system, in turn, has two branches that are interrelated. These are the humoral and cell-mediated or cellular immune responses. The humoral response is generally mediated by immunoglobulins produced by B cells or B lymphocytes. In turn, it activates the cellular response which involves the T cells or T lymphocytes that are programmed with information to detect surface molecules specific to the pathogen (Spiering, 2015).

There is also the so-called complement system, which consists of several plasma proteins that react with one another to opsonize (or make them susceptible to phagocytosis) pathogens and triggers a series of inflammatory responses to fight infection. It was initially thought to be a "complement" or effector of the antibacterial activity of immunoglobulins. However, it was soon observed that it can also be activated without immunoglobulins. Now it is known to play an important role in the innate immune system. This can be activated via three distinct pathways, namely the classical pathway, alternative pathway, and the mannose or mannan-binding lectin (MBL) pathway (Janeway *et al.*, 2001).

CHAPTER REFERENCES

1. Institute for Quality and Efficiency in Health Care (2020). *How does the immune system work?* https://www.ncbi.nlm.nih.gov/books/NBK279364/ (Accessed on 23 April 2021).
2. Janeway, C.A., Jr., Travers, P., Walport, M. & Shlomchik, M.J. (2001). Immunobiology: The Immune System in Health and Disease (5th edn.). New York: Garland Science.https://www.ncbi.nlm.nih.gov/books/NBK27156/ (Accessed on 17 May 2021).
3. Lacy, P. & Stow, J. (2011). Cytokine release from innate immune cells: association with diverse membrane trafficking pathways. *Blood* 118(1): 9-18. https://doi.org/10.1182/blood-2010-08-265892 (Accessed on 17 May 2021).
4. Marshall, J.S., Warrington, R., Watson, W. & Kim. H.L. (2018). An introduction to immunology and immunopathology. *Allergy, Asthma & Clinical Immunology* 14:49. https://doi.org/10.1186/s13223-018-0278-1 (Accessed on 26 April 2021).
5. McComb, S., Thiriot, A., Akache, B., Krishnan, L. & Stark, F. (2019). Introduction to the Immune System. In: *Immunoproteomics. Methods in Molecular Biology, vol. 2024*; Fulton, K. & Twine, S. (Eds.). New York: Humana. pp.1-24.
6. National Cancer Institute, 2021. *Reactive oxygen species*. https://www.cancer.gov/publications/dictionaries/cancer-terms/def/reactive-oxygen-species (Accessed on 17 May 2021).
7. Oldham, K. (2021). Chemokines: Introduction. *British Society for Immunology*.https://www.immunology.org/public-information/bitesized-immunology/receptors-and-molecules/chemokines-introduction (Accessed on 17 May 2021).
8. Spiering, M.J. (2015). Primer on the Immune System. *Alcohol Research* 37(2):171-175.
9. Testar, J. (2021). Cytokines: Introduction. *British Society for Immunology*. https://www.immunology.org/public-information/bitesized-immunology/receptors-and-molecules/cytokines-introduction (Accessed on 17 May 2021).

FIGURE REFERENCES

1. Opening image. Brinkmann, V. (2005). PLoS Pathogens 1(3): Cover page. Creative commons attribution 2.5 generic license; https://commons.wikimedia.org/wiki/File:Immune_response.svg (Accessed on 23 April 2021).
2. Figure 1.1. Vassiliev, N. (2018). Creative commons attribution-sharealike 4.0 international license; https://commons.wikimedia.org/wiki/File:Immune_response.svg (Accessed on 23 April 2021).
3. Figure 1.2. Spiering, M.J. (2015). Primer on the Immune System. Alcohol Research 37(2):171-175. Public domain.

CHAPTER 2

INNATE IMMUNE SYSTEM

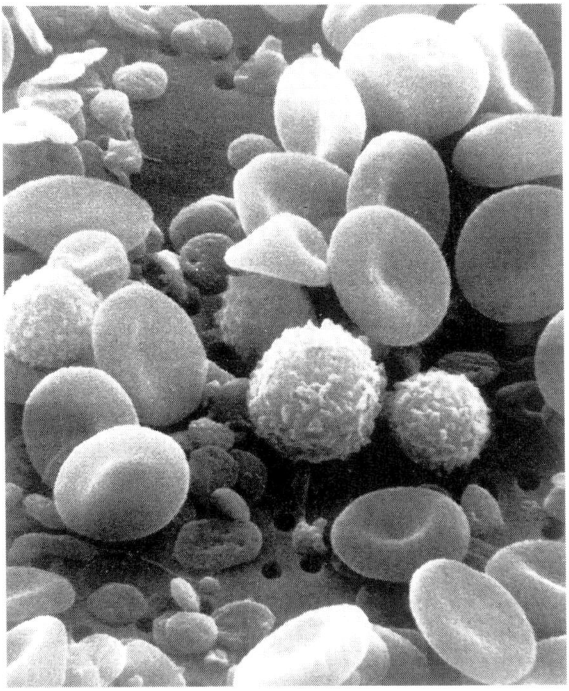

(Source: Bruce Wetzel & Harry Schaefer, Public Domain)

Countless potential pathogens enter the body through contact, ingestion, and inhalation. If a particular pathogen has been encountered before, then infection is avoided due to the action of the adaptive immune system. For new pathogens, the body relies on the innate immune system during the first critical hours and days of exposure (Alberts *et al*., 2002). Therefore, it serves as the first line of defense during infection and triggers an inflammatory response to invaders (Mogensen, 2009).

Innate immunity is not pathogen- or antigen-specific. Still, it demonstrates a degree of specificity,and it can differentiate between foreign and self-antigens. It relies on phagocytic cells, antigen presenting cells (APCs), and proteins that detect conserved features of pathogens and becomes rapidly activated to destroy invaders. Innate immunity is also interconnected with adaptive immunity since it is involved in activating and shaping the adaptive immune response (Alberts *et al*., 2002; Mogensen, 2009). In summary, the innate immune system has a critical role in preventing the entry of pathogens into the body through physical and chemical barriers, in avoiding the spread of infections through the complement system and other humoral factors, in removing pathogens through phagocytosis and cytotoxicity mechanisms, and in activating the adaptive immune system through antigen presentation and cytokine production (Muñoz-Carrillo *et al*., 2017).

2.1 CELLS OF THE INNATE IMMUNE SYSTEM

Some cells involved in the innate immune response form physical barriers that keep out pathogens. Many of them are phagocytes, which engulf other cells or molecules via a process known as phagocytosis. A more detailed discussion of this process is found in Section 2.2.3. The ingested harmful microorganisms or foreign particles are then inactivated by being rapidly bombarded with reactive oxygen species (ROS), in a process known as respiratory or oxidative burst (Spiering, 2015). ROS refers to a type of unstable molecule that contains oxygen and that easily reacts with other molecules in a cell. These are free radicals and hence, are also called oxygen radicals. The accumulation of ROSmay damage the DNA, RNA, and protein components of cells possibly leading to cell death (National Cancer Institute, 2021).

Other cells involved in this type of immune response may neutralize pathogens by producing antimicrobial compounds. Several of these cells have immunomodulatory functions by recognizing certain structures in various pathogens resulting in the production of cytokines.

2.1.1 Non-myeloid Cells

Non-myeloid cells refer to cells whose development does not take place in the bone marrow. They form a barrier between the internal and external environment. These produce antimicrobial substances called antimicrobial peptides (AMPs) that prevent the entry of pathogens into the body. AMPs in humans are subdivided into two families, namely defensins and cathelicidins. Defensins and cathelicidins are first synthesized in their inactive, prepropeptide form consisting of an N-terminal signal sequence for targeting to the endoplasmic reticulum (ER), a pro segment, and a C-terminal signal sequence that has antimicrobial activity after it is cleaved during the later stages of intracellular processing or after secretion (Bals & Hiemstra, 2004). The cleaved and mature form of cathelicidin is called LL37. Defensins are further classified into α, β, and θ types. Cathelicidins and defensins act against gram-positive and gram-negative bacteria. These are positively charged and interact electrostatically with the negative charge of the cell wall or cell membrane of certain pathogens causing the formation of pore-like structures in the membrane and ultimately cell lysis. They can also inhibit protein and DNA synthesis in bacteria. They also enable cytokine release, cell proliferation, chemotaxis, angiogenesis, and wound healing (Muñoz-Carrillo *et al.*, 2017; Coates *et al.*, 2018; Okumura & Takeda, 2017).

2.1.2 Myeloid Cells: Granulocytes

Myeloid or myelogenous cells come from hematopoietic stem cells in the bone marrow. These include the white blood cells or leukocytes. Leukocytes with granules in their cytoplasm are called granulocytes. Examples of these are neutrophils, basophils, and eosinophils.

2.1.2.1 Neutrophils

The most abundant granulocytes are the neutrophils, and it is the most abundant leukocyte with the greatest bactericidal activity. Blood levels of these cells can reach up to 5 million cells per milliliter (mL) of blood comprising 60-70% of circulating leukocytes. They normally circulate in the blood. Owing to their segmented nucleus, neutrophils are also called polymorphonuclear leukocytes (see Figure 2.1a). Upon being induced by cytokines and chemokines, they quickly migrate to the site of injury or infection so that they are among the first immune cells to arrive at the affected area. Neutrophils detect pathogens via Toll-like receptors (TLRs) and then ingest them by phagocytosis forming so-called phagosomes, which are vacuoles containing the ingested particle (see Figure 2.1b,c)(Spiering, 2015). They also form and release to the extracellular space neutrophil extracellular traps (NETS). These consist of DNA fibers that sequester and kill pathogens. They can also contribute to inflammatory response and their short lifespan leads to fast resolution of inflammation. They promote wound healing by stimulating local tissue remodeling and macrophage recruitment (Sheshachalam *et al.*, 2014). Neutrophils also provide signals for the activation and maturation of macrophages and dendritic cells. Also, they are involved in regulating T cell-mediated immune response. These indicate that neutrophils may also play a role in adaptive immunity (Kumar & Sharma, 2010).

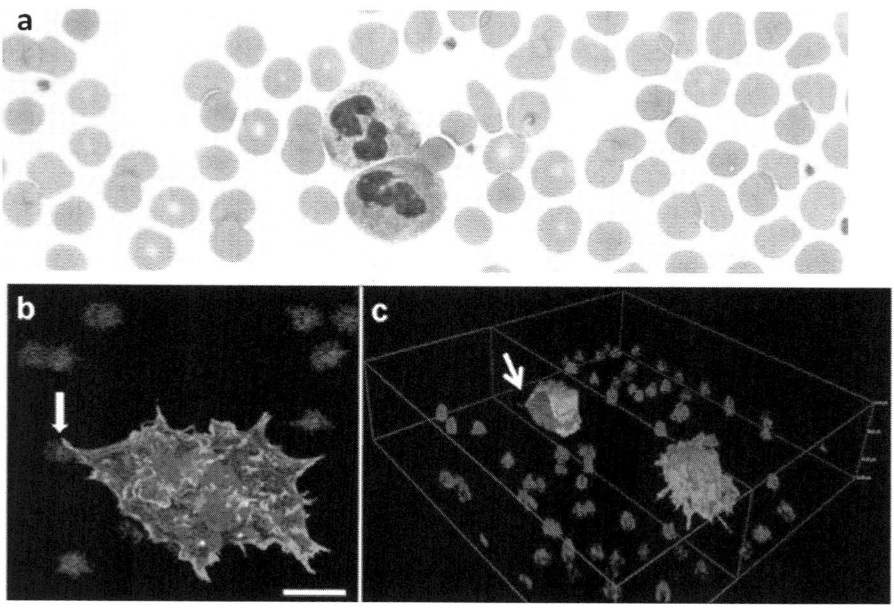

Figure 2.1. (a) A neutrophil is a kind of leukocyte with granular cytoplasm and segmented nucleus. It can function as a phagocyte in which it can (b) extend its membrane (green) to come in contact with foreign particles such as bacteria (red) and (c) engulfs them then subsequently degrades them. (Source: a. Graham Beards, Creative Commons License; b. María Lázaro-Díez *et al.*, Creative Commons License)

As previously mentioned, granulocytes like neutrophils can neutralize pathogens by employing oxidative burst. This is an oxygen-dependent bactericidal activity whereby NADPH oxidase produces superoxide radicals, hydrogen peroxide, hypochlorous acid, hydroxyl radicals, and chloramines. These are then converted to ROS that effectively kill bacteria. Neutrophils also employ oxygen-independent bactericidal process by producing a mixture of DNA and AMPs and proteins that trap and kill microbes. These molecules are contained within cytoplasmic granules and are delivered to the phagosome, which contains the ingested bacterium after phagocytosis. In essence, phagocytosis triggers ROS production and fusion of cytoplasmic granules with phagosomes resulting in the potent antimicrobial activity of neutrophils (Kobayashi *et al.*, 2018).

Neutrophils have at least three types of granules. These are the primary or azurophilic, secondary or specific, and tertiary or gelatinase granules. Azurophilic granules contain potent antimicrobial peptides or proteins such as α-defensins, cathepsins, proteinase-3, elastase, azurocidin or CAP37, and lysozymes (Sheshachalam *et al.*, 2014; Kobayashi *et al.*, 2018). The α-defensins are relatively small cationic polypeptides (3-5 kiloDaltons/kDa), which interact with the negatively-charged molecules at the surface of the pathogen such as the anionic lipids (e.g., phosphatidylglycerol and cardiolipins) and other anionic components (e.g., lipopolysaccharide and lipoteichoic acid) of their cell membrane. These cationic AMPs can thus insert themselves into the membrane so that membrane integrity is disrupted,and the bacterium undergoes osmotic lysis. AMPs are also involved in chemotaxis, wound repair, and stimulation of histamine release so that they serve as an important bridge between innate and adaptive immune responses. Cathepsins, proteinase-3, elastase, and azurocidin are collectively called serprocidins, a family of antimicrobial proteins that are structurally homologous with serine proteases. In fact, with the exception of azurocidin, these proteins are serine proteases with direct antimicrobial activity. Specifically, elastase can cleave the outer membrane protein A (OmpA) of *Escherichia coli* leading to membrane disruption and eventually cell death. On the other hand, lysozyme can degrade bacterial peptidoglycan and can be found in other types of granules. Secondary granules are rich in lactoferrin, which binds and sequesters iron so that it cannot be used for bacterial growth but supplies it to produce neutrophil hydroxy radicals (Kobayashi *et al.*, 2018). Meanwhile, tertiary granules contain matrix metalloproteinases (MMPs), which are necessary for degrading extracellular matrix materials so that cells can migrate through tissues during tissue reorganization (Sheshachalam *et al.*, 2014; Shamri *et al.*, 2011). The fourth type of granules was recently described. These are found to be enriched in the microbial lectin, ficolin-1. Ficolin-1 is found in tertiary granules, but these potentially fourth types of granules are ficolin-1-rich and gelatinase-poor. These have a high exocytosis rate, which allows the rapid release of pattern recognition molecules to activate the lectin complement pathway (Sheshachalam *et al.*, 2014).

2.1.2.2 Basophils

Basophils are polymorphonucleate leukocytes which are called such because their granules stain with basophilic dyes (see Figure 2.2). They are the least abundant leukocyte circulating in the blood, constituting $\leq 1\%$ of the leukocyte population. These are produced from the bone marrow and are usually present in the peripheral blood and in the spleen but can also enter the lymph nodes and tissue inflammatory sites. They function as immune regulatory cells and as effector immune cells. They are recruited to inflamed tissues by the upregulation of chemokine receptors on their membranes in response to inflammatory chemokines. In the site of inflammation, the activated basophils release serine proteases and enhance the local micro-vascular permeability to allow T-cells and innate immune cells to reach the area. Basophils can sometimes cause inflammatory responses such as allergic reactions. In fact, basophils, along with eosinophils, are major effectors of allergy among granulocytes. Basophils are morphologically similar to mast cells and both of them express immunoglobulin E (IgE), which mediates allergic reactions, on their membranes. Both of these immune cells also release histamine and leukotrienes after stimulation with allergens and IgE-dependent triggers (Spiering, 2015; Chirumbolo *et al.*, 2018). They also produce proallergic cytokines like interleukin 4 and 13 (IL4 and IL13). In fact, basophils are responsible for producing 72% of IL4 in the bronchial mucosa and for producing the early batch of IL4 and IL13 in the peripheral blood during allergen-induced asthmatic reactions (Gibbs, 2008). Basophils also produce heparin, which is an anticoagulant or a substance that reduces blood clotting, as well as serotonin, which contributes to wound swelling (Spiering, 2015).

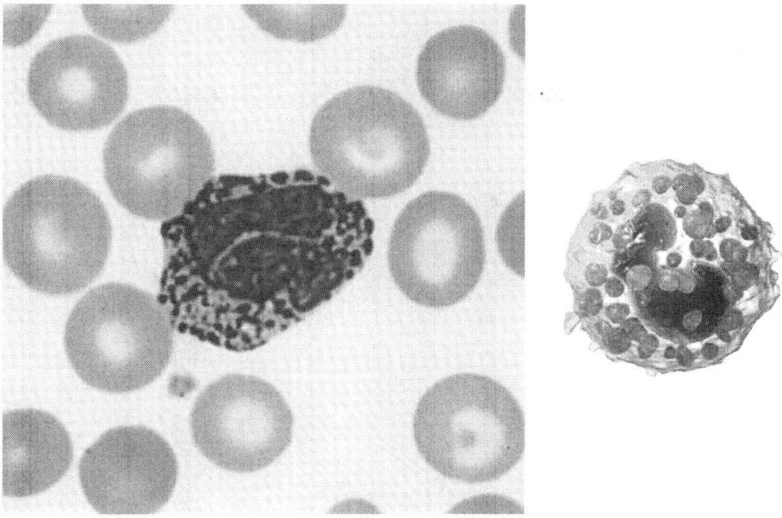

Figure 2.2. A basophil is polymorphonucleated with cytoplasmic granules that have an affinity with basophilic dyes. (Source: Left-A. Rad *et al*., Creative Commons License; Right-Bruce Blaus& Blausen.com staff, Creative Commons License)

Basophils are also involved in the removal of parasitic helminths or nematodes (worms)since they release IL4 and IL13 and respond to cytokines like IL5, IL9, IL18, and IL33(Chirumbolo *et al*., 2018). IL4 production during nematode infection promotes the differentiation of naïve CD4 (cluster of differentiation 4)T cells into type 2 helper T(Th2) cells (Falcone *et al*., 2000). Cluster of differentiation is a designation to approximately 250-400 different antigens or proteins on the surface of leukocytes. Each unique surface molecule is assigned a different number as a way to identify the phenotype of leukocytes. CD molecules often act as receptors or ligands. Some are involved in cell signaling while others mediate cell adhesion (Xiong & Xu, 2014; Aktor, 2019).

Basophils are important for the expulsion of parasites from the large intestine. In the presence of IgE, antigens from the eggs of the parasite *Schistosoma mansoni* induce basophils to release IL4 and other mediators *in vitro*. Additionally, the hookworm *Necator americanus* produces proteases that induce the basophilic cell line to synthesize IL4 and IL13 (Falcone *et al*., 2000).

In mouse models, basophils were also found to release IL6, which induces Th17 cell differentiation and immune response. In humans, they are possibly involved in Th17-mediated immunity in lung and bowel inflammatory diseases, where they may have a role in the increase of Th17/ Th1 cytokine expression in memory CD4 T cells (Chirumbolo *et al.*, 2018).

2.1.2.3 Eosinophils

Eosinophils are granulocytes with bilobed nuclei,and they stain pink with eosin (see Figure 2.3). They are found in relatively low numbers in circulating blood, i.e., 0.5-1% of total leukocytes in normal individuals but this percentage can increase up to 3-5% during an allergic reaction and could even be higher during parasitic infection. Under healthy conditions, the majority of the eosinophils are located within mucosal tissues of the gastrointestinal and genitourinary tracts and within primary and secondary lymphoid tissues. They contain in their cytoplasm specific granules, which are also called secondary or crystalloid granules. These granules are made up mostly of cytotoxic cationic proteins, and also store a wide variety of cytokines, chemokines, and growth factors, which can be immediately released upon stimulation. This ability of eosinophils to secrete mediators without the need for *de novo* protein synthesis differentiates them from many other innate and adaptive immune cells. Eosinophils can sense pathogens and promote innate immunity because they express a wide range of receptors that allow them to respond to many types of cytokines, chemokines, and lipid mediators (Buckland, 2021; Shamri *et al.*, 2011).

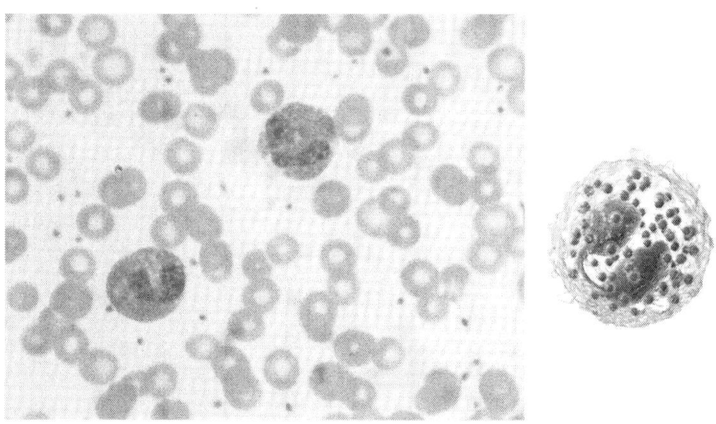

Figure 2.3. An eosinophil is a granulocyte with a bilobed nucleus and with cytoplasmic granules that stain pink with eosin dye. (Source: Left- Ed Uthman, Creative Commons License; Right- Bruce Blaus & Blausen.com staff, Creative Commons License)

They are known for their cytotoxic effector functions aimed at parasites and at host tissues in allergic diseases. They are also involved in regulating inflammation, maintaining epithelial barrier function, affecting tissue remodeling, and bridging innate and adaptive immunity. Phagocytosis was also observed among eosinophils, but it is less efficient than neutrophil phagocytosis. Also, eosinophils function as professional antigen presenting cells (APCs) (Shamri *et al.*, 2011). This means that they can attach antigens or pieces of the pathogen to their cell surface and present these antigens from viruses, bacteria, and parasites to CD4 T cells (Travers & Rothenberg, 2015).

Moreover, eosinophils are known to have immunoregulatory functions. For instance, eosinophils promote differentiation of naïve CD4 T cells into Th2 cells by releasing IL4, IL25, and indoleamine 2,3-dioxygenase. They are also associated with the effector functions of Th2 cells in allergic diseases and helminth parasite infections. They promote Th2 cell recruitment by inducing the expression of chemokines that serve as T-cell chemoattractants and by enhancing cytokine production by T cells. They interact with other immune cells such as mast cells, dendritic cells, and B lymphocytes. It was also exhibited *in vitro* that eosinophils promote B cell proliferation, survival, and antibody production. The IL4 secreted by eosinophils induces B lymphocyte differentiation leading to the production of IgM-producing plasma cells (Shamri *et al.*, 2011; Travers & Rothenberg, 2015).

Cationic proteins that reside in eosinophils include major basic proteins (MBP), eosinophil peroxidase (EPO), and eosinophil-associated RNases (EARs) such as eosinophilic cationic protein (ECP) and eosinophil-derived neurotoxin (EDN). Residual amounts of MBP, ECP, and EDN are also found in other granulocytes like neutrophils, basophils, and activated macrophages. These cationic proteins have antibacterial, antifungal, antiparasitic, and tissue cytotoxic properties. EDN was also shown to have antiviral activity *in vitro*. These cationic proteins also have mediating functions through signaling interaction with other cells (Shamri *et al.*, 2011). For instance, EDN is a known chemoattractant of dendritic cells while EPO induces dendritic cells to migrate to draining lymph nodes. Additionally, both EDN and EPO induce dendritic cell activation and maturation (Travers & Rothenberg, 2015).

The eosinophilic granules also contain the chemokines like eotaxin; cytokines, which include IL2, IL4, IL5, IL6, IL10, IL12, IL13, IL25, interferon-gamma (IFNγ), and tumor necrosis factor-alpha (TNFα); and growth factors like granulocyte macrophage colony-stimulating factor (GM-

CSF), transforming growth factor-beta (TGFβ), and proangiogenic factors that promote angiogenesis or the generation of new blood vessels. Eosinophils also produce lipid mediators such as cysteinyl leukotrienes and arachidonic-based inflammatory mediators. They also express matrix metalloproteinases like the soluble MMP9 and the membrane-bound MMP17. They store other enzymes like acid phosphatase, collagenase,arylsulfatase B, histaminase, phospholipase D, catalase, and non-specific esterases (Shamri *et al.*, 2011).

Eosinophils exhibit a unique antibacterial mechanism that is induced by the encounter of IL5- or IFNγ-primed eosinophils with Gram-negative lipopolysaccharide (LPS). This involves the release of extracellular traps from eosinophils. These traps contain mitochondrial DNA (mtDNA) and cytotoxic granule-derived proteins. This process does not affect the eosinophil's viability. In contrast, similar processes that occur in neutrophils and mast cells result in cell death (Shamri *et al.*, 2011).

2.1.2.4 Mast cells

Mast cells are long-lived cells that can survive for months or even years. They are involved in both innate and adaptive immunity. They do not usually circulate in the blood but are found in connective tissues all over the body, especially under the skin, near blood vessels in all tissues except for the central nervous system and the retina, near lymph vessels, in nerves as well as in mucous membranes where they assist in wound healing and in defending against pathogens. The connective tissue mast cells produce chymase, tryptase, and carboxypeptidase, whereas the mucosal mast cells produce tryptase only. In tissue sections, mast cells are distinguished by staining blue with Toluidine blue stain and other blue dyes (see Figure 2.4a). They contain 50-200 large granules in their cytoplasm (see Figure 2.4b), which contain the majority of the body's histamine. These large granules served as the basis of their name, "Mastzellen", which is German for well-fed cells. Pathogens, allergens, or even injury can activate these cells to release their granules, which apart from histamine, are also rich in heparin, cytokines, and growth factors. These promote inflammation during allergic reactions and immune responses against parasites. Their granules also store chondroitin sulfate and neutral proteases. Mast cells also function to regulate vasodilation, vascular homeostasis, angiogenesis, and venom detoxification (Spiering, 2015; National Cancer Institute, 2021; Krystel-Whittermore *et al.*, 2016; Amin, 2012).

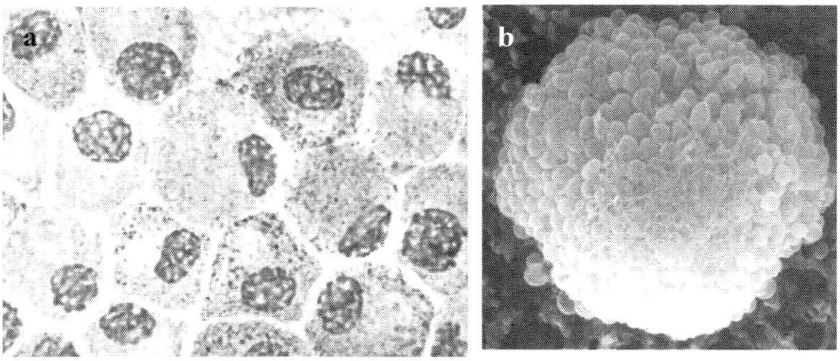

Figure 2.4. (A) Mast cells appear blue when stained with Toluidine blue. These cells appear slightly degranulated because they were activated by an antigen. (Magnified 100X). (B) A scanning electron micrograph of rat peritoneal mast cells showing the numerous granules. (Source: a-Kauczuk, Creative Commons License; b-Melissa Krystel-Whittermore *et al.*, Creative Commons License)

Just like leukocytes, they are derived from the bone marrow but unlike them, mast cells are released into the blood as mast cell progenitors and then they are recruited into certain tissues where they undergo differentiation and proliferation through the influence of the stem cell factor binding to the c-kit receptor and other growth factors present in the tissue (Weller, 2021; Krystel-Whittermore *et al.*, 2016). These progenitor cells express CXCR2, which directs their migration to the small intestine. As a result, there is a great quantity of mast cell progenitors in the small intestine. In contrast, there are no mast cell progenitors in the lungs under normal physiological conditions. They are recruited into the respiratory endothelium when there is antigen-induced inflammation. When mature mast cells are activated and degranulated, more mast cell progenitors are recruited to the site of inflammation (Krystel-Whittermore *et al.*, 2016).

As an innate immunity cell, mast cells bind antigens by means of toll-like receptors (TLRs) and receptors for the complement system. This induces the release of inflammatory mediators. Lipopolysaccharide of Gram-negative bacteria is bound by TLR4 causing the release of pro-inflammatory cytokines, such as TNFα, IL1, and IL6, without degranulation. Gram-positive bacteria, and to an extent Gram-negative bacteria and mycobacteria, activate TLR2 of mast cells leading to the release of cytokines like IL4. The release of inflammatory mediators facilitates the elimination of bacteria because these increase vascular permeability, increase fluid accumulation, and recruit other immune cells like eosinophils, natural killer (NK) cells, and neutrophils. Moreover, mast cells produce antibacterial compounds, like cathelicidins, defensins, and psidins. In the case of viral infection, mast cells recruit CD8 T cells, which produce IFNα and IFNβ. Mast cell degranulation and histamine release are stimulated by TLR2 activation by the peptidoglycan of Gram-positive bacteria. Furthermore, IgE can activate mast cells to release mediators that increase vascular permeability and smooth muscle contraction. This induces vomiting or diarrhea that helps expel parasites from the gastrointestinal tract. It also induces coughing in order to remove parasites in the respiratory tract (Krystel-Whittermore *et al*., 2016).

In terms of its involvement in adaptive immunity, mast cells can process and present antigens via class I and class II major histocompatibility complex (MHCI and MHCII). MHC molecules bind peptide fragments derived from pathogens and display them on the cell surface for recognition by the appropriate T cells. This results in the destruction of virus-infected cells, activation of macrophages to kill bacteria inside intracellular vesicles, and activation of B cells to produce antibodies (Janeway *et al*., 2001). Mast cells can also activate dendritic cells, which are also capable of antigen presentation, and cytotoxic T cells (Krystel-Whittermore *et al*., 2016).

2.1.3 Myeloid Cells: Agranulocytes

Agranulocytes are leukocytes without distinct granules in their cytoplasm. These include lymphocytes and monocytes.

2.1.3.1 Lymphocytes

Lymphocytes that are small, 7-10 μm in diameter, have a nucleus that stains dark purple with Wright staining and they have less cytoplasm (see Figure 2.5a). They are usually naïve or not yet stimulated by antigens. Upon stimulation, they become large and granular measuring 10-12 μm in diameter and they contain more cytoplasm, organelles, RNA, and scattered granules (Lewis & Blutt, 2019; Bezuidenhout & Schneider, 2009) (see Figure 2.5b). Lymphocytes constitute 25% of leukocytes in the blood. There are three types of lymphocytes, namely T cells, B cells, and innate lymphoid cells (ILCs). They make up 80%, 10%, and 10% of the total blood lymphocyte population, respectively. Lymphocytes are also found in the lymph and lymph organs, such as the thymus, lymph nodes, spleen, and appendix. T cells make up 90% of lymphocytes in the thymus while they constitute approximately 30-40% of lymphocytes in the spleen and lymph nodes. B cells make up the greater percentage, i.e., 60-70%, of lymphocytes in these tissues (Lewis & Blutt, 2019).

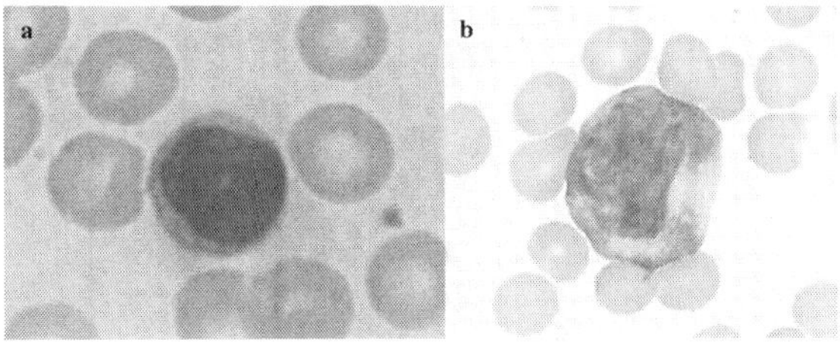

Figure 2.5. (a)A small, unstimulated lymphocyte has little amount of cytoplasm with purple-staining nucleus. (b) A large, stimulated one becomes granular with more cytoplasm. (Source: a-Nicolas Grandjean, Creative Commons License; b-SpicyMilkBoy, Creative Commons License)

T cells directly attack pathogens, and they also produce cytokines. On the other hand, B cells produce antibodies or immunoglobulins, which attach to foreign invaders triggering their eventual destruction (Brody, 2021). Both T and B cells are involved in adaptive immunity and so their specific roles are discussed further in the next chapter. The third type, ILCs, as their name implies, are involved in innate immune responses. However, unlike B and T cells, they do not have antigen-specific receptors. It is likely that adaptive and innate lymphocytes work cooperatively in order to provide a strong immune response against microbes. There are several types of ILCs, such as classical natural killer (cNK) cells, lymphoid tissue inducer (LTi) cells, thymic natural killer (NK) cells, NK receptor-positive (NKR+) **LTi** cells, and Th2-type ILCs in mice like natural helper (NH) cells and nuocytes. NK cells, LTi cells, and Th2-type ILCs produce Th1, Th17, and Th2 cytokines, respectively (Koyasu & Moro, 2012).

ILCs limit the replication of pathogenic microorganisms. In the case of viral infection, cNK cells have certain receptors such as CD94/natural killer group 2 member D (NKG2D) that recognize antigens on target cells that are infected. Then the cNK cells limit viral propagation by causing the lysis of infected cells. Their cytotoxic activity is due to various effector molecules such as perforin, granzyme, Fas ligand (Fas-L), and TNF-related apoptosis-inducing ligand (TRAIL). Adult LTi cells produce IL17 and IL22. IL22 that is produced by NKR+LTi cells induces the epithelial cells to produce antimicrobial peptides (AMPs) like β-defensin psoriasin, calgranulin A, calgranulin B, regenerating islet-derived III beta protein (RegIIIβ), and RegIIIγ. Therefore, IL22 is critical in providing innate immune protection against specific pathogens such as *Citrobacter rodentium*, *Klebsiella pneumoniae*, and *Candida albicans* (Koyasu & Moro, 2012).

On the other hand, Th2-type ILCs in mice play a role in providing protection against parasites since Th2 activity is essential for anti-helminth immunity. Th2 response is also involved in antigen-specific allergic responses. One type of Th2-type ILC is the NH cell. NH cells were discovered in adipose tissues and are found in lymphoid clusters called fat-associated lymphoid clusters (FALC). They provide protection against helminth infection, possibly by producing IL5 and IL13 during the early phase of helminth infection. These cytokines induce eosinophilia (i.e., higher than normal eosinophil levels) and goblet cell hyperplasia, which are necessary for Th2-mediated expulsion of helminths. In contrast to NH cells, nuocytes are found in lymph nodes and express chemokine receptors that enable them to be recruited to the lymph nodes. They produce IL4. Lymphocytes similar to mouse NH have been observed in the human gut, lung, mesentery, and blood (Koyasu & Moro, 2012).

2.1.3.2 Monocytes

Monocytes are the largest leukocyte, 12-20 μm in diameter, with a bilobed or kidney-shaped nucleus (see Figure 2.6). They have a one to three-day half-life. They circulate in the blood making up about 5% of the circulating leukocytes in the blood. They are also stored in the spleen and lungs ready to be mobilized when needed. They can perform phagocytosis, present antigens, produce chemokines, and proliferate upon infection and injury. The phagocytic and antigen-presenting functions of monocytes enable them to ingest and remove microorganisms, foreign material, and dead or damaged cells. Monocytes have toll-like receptors (TLRs) that detect invading microbial cells. In response to such stimuli as well as to chemokine mediators, they move from the bone marrow into the bloodstream and infiltrate infected tissues within 12 to 24 hours (Espinoza & Emmady, 2020; Chiu & Bharat, 2016).

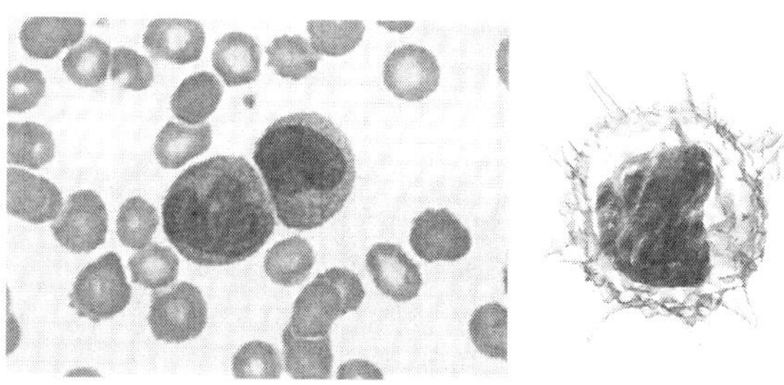

Figure 2.6. Monocytes are the largest among the leukocytes with bilobed nuclei. (Source: Left-Graham Beards, Creative Commons License; Right-Bruce Blaus & Blausen.com staff, Creative Commons License)

Monocytes are derived from myeloid precursor cells in primary and secondary lymphoid organs, like the liver and bone marrow (see Figure 2.7). In the peripheral blood, monocytes respond to signals from the microenvironment enabling them to differentiate into three types, namely classical inflammatory monocytes, non-classical endothelial patrolling monocytes, and intermediate monocytes. These differ in terms of their CD14 and CD16 expression profile. Classical monotypes are $CD14^{++}CD16^{-}$, non-classical ones are $CD14^{+}CD16^{++}$, and intermediate monocytes are $CD14^{++}CD16^{+}$. Monocytes can migrate transendothelially via the interaction of α4β1 integrin with vascular cell adhesion molecule 1(VCAM-1) (van den Bosch *et al.*, 2017).

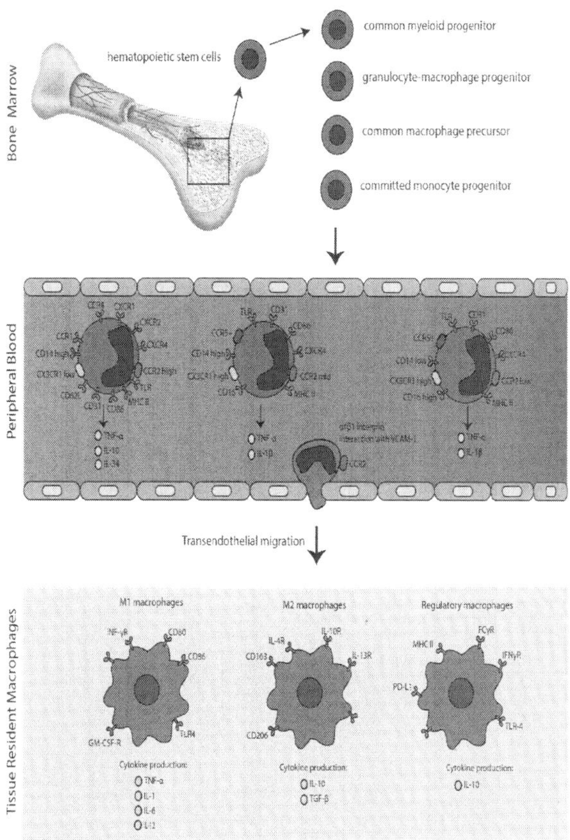

Figure 2.7. Monocytes originate from progenitor cells (i.e., common myeloid progenitor cells, granulocyte-macrophage progenitor cells, common macrophage precursor cells, committed monocyte progenitor cells) in lymphoid organs

like the bone marrow. In the peripheral blood, the monocytes have the capacity to differentiate into three subsets that have distinct functions and phenotypes in terms of CD14 and CD16 expression. Additionally, monocytes can also undergo transendothelial migration and subsequently differentiate into macrophages, which are activated and polarized to acquire pro-inflammatory (M1), anti-inflammatory (M2), or the regulatory (Mreg) phenotype. (Source:Thierry P.P.van den Bosch*et al.*, Creative Commons License)

Classical monocytes proliferate in the bone marrow in response to infection or injury. During bacterial infection, they are released into the blood circulation and proceed to the site of infection. They will then phagocytose the pathogens, secrete certain chemokines that recruit other immune cells, and present the antigen through MHC II (Chiu & Bharat, 2016). The classical monocytes stored in the spleen can also be mobilized to the site of injury in response to certain signals, such as angiotensin II during myocardial infarction. These cells can also differentiate into dendritic cells and macrophages(Chiu & Bharat, 2016).

Classical monocytes that returned to the bone marrow possibly mature into non-classical endothelial patrolling monocytes. These are motile with a distinct crawling pattern as they move along the luminal side of the endothelium. Their crawling motion is independent, frequently going against, the direction of blood flow, but is dependent on integrin. They perform phagocytosis of the injured endothelium and recruit neutrophils to the site of injury. Non-classical monocytes differentiate into dendritic cells in the lungs. They have been observed to limit metastases of lung tumors (Chiu & Bharat, 2016). These cells also promote wound healing and angiogenesis in models of atherosclerosis and cardiac infarction (De Kleer *et al.*, 2014).

Non-classical monocytes secrete cytokines, but there are conflicting reports on whether they produce pro-inflammatory or anti-inflammatory cytokines. Studies to resolve thisissue led to the discovery of intermediate monocytes. This type of monocyte is a transitional population between classical and non-classical types and expresses higher levels of MHC II (Chiu & Bharat, 2016).

Meanwhile, macrophages are found in lymphoid and non-lymphoid tissues, responding to pathogens and injury by performing phagocytosis (see Figure 2.8) and producing inflammatory cytokines. They also have a critical role during development and in tissue homeostasis, since they clear cell debris and produce growth factors (Chiu & Bharat, 2016; De Kleer *et al.*, 2014). There are two populations of macrophages, the resident and the inflammatory ones. Resident macrophages occur in almost all tissues and contribute to their development and repair as well as immunological surveillance. Inflammatory macrophages come from circulating monocytes and rapidly infiltrate injured or infected tissues (Muñoz-Carrillo *et al.*, 2017).

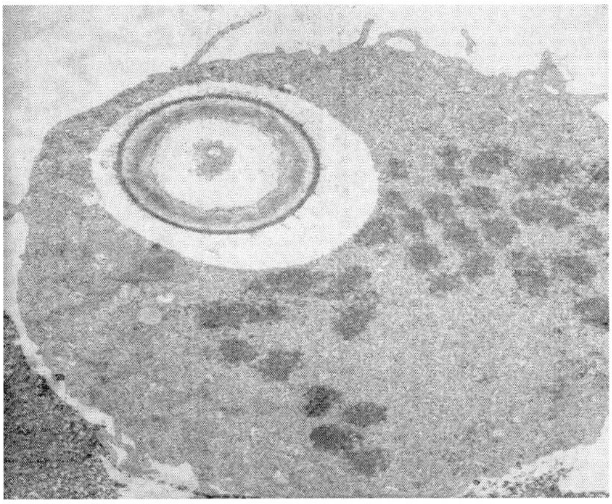

Figure 2.8. Macrophages are phagocytes. In this case, a macrophage engulfed a fungal pathogen (pale yellow circle). (Source: Carolina Coelho, Creative Commons License)

In the early embryonic stage of mammalian development, early monocyte progenitors can move to different organs like the liver, skin, and lungs and differentiate into resident macrophages that will self-maintain throughout life. However, not all macrophages are monocyte-derived (see Figure 2.9). Embryonic macrophages derived from myeloid progenitors in the yolk sac migrate to the central nervous system and form microglia cells, which are maintained without being replenished by adult hematopoiesis. Yolk sac macrophages also contribute to adult tissue macrophage pools such as the Kupffer cells in the liver. On the other hand, the skin-resident macrophages, epidermal Langerhans cells, arise from both yolk sac macrophages and fetal liver-derived monocytes. Meanwhile, alveolar macrophages in the lungs are completely derived from fetal liver-derived monocytes. These tissue-resident macrophages are capable of self-renewal. The macrophages in the intestinal lamina propria are replenished by fetal liver and bone marrow-derived monocytes. When $CD14^+$ monocytes are recruited to sites of injury or inflammation, they can differentiate into inflammatory macrophages and monocyte-derived dendritic cells (MoDCs) (van den Bosch *et al.*, 2017; De Kleer *et al.*, 2014).

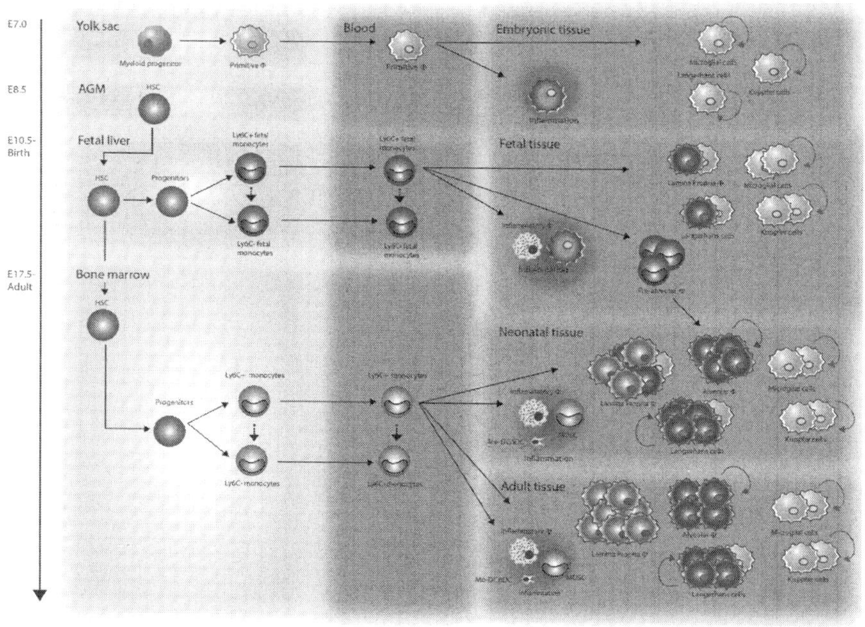

Figure 2.9. Tissue-resident macrophages are produced at different stages of development and arise from three different sources. In the early stage of embryogenesis, myeloid progenitors from the yolk sac give rise to microglia in the brain, Kupffer cells in the liver, and Langerhans cells in the skin. When hematopoiesis starts to occur in the liver of the fetus, fetal liver-derived monocytes differentiate and contribute to the pool of Langerhans cells in the skin and macrophages in the gut lamina propria. They also seed the lung with pre-alveolar macrophages before birth. After birth, these rapidly differentiate into long-lived alveolar macrophages. Unlike the other tissue macrophages that proliferate after birth, the lamina propria macrophages are continuously renewed through the differentiation of bone marrow-derived monocytes. Monocytes can also be recruited to sites of infection or injury and differentiate into inflammatory macrophages, monocyte-derived dendritic cells (Mo-DCs), or myeloid-derived suppressor cells (MDSCs). (Source: Ismé De Kleer *et al.*, Creative Commons License)

Macrophages are activated in two ways (see Figure 2.7). The first is known as classical or M1 activation, which responds to IFNγ, bacterial lipopolysaccharide or TNFα. M1 macrophages are induced by pathogens and tumor cells to increase the expression of nitric oxide synthase and to secrete pro-inflammatory chemokines and cytokines, like IL1, IL6, IL8,

IL12, C-C motif chemokine ligand 2 (CCL2), C-X motif chemokine ligand 9 (CXCL9), CXCL10, and TNFα (Chiu & Bharat, 2016; Aktor, 2012; Muñoz-Carrillo *et al.*, 2017). They also present antigens via MHC II leading to initiation of inflammation, recruitment of granulocytes, and promotion of Type 1 helper T (Th1) cell response. The alternative or M2 activation results to a more varied immune response. M2-activated macrophages are linked to Th2 cell response, downregulation of the initial inflammatory response and promotion of the resolution of inflammation, initiation of tissue healing and fibrosis, and tumor progression (Chiu & Bharat, 2016; Germic *et al.*, 2019). M2 macrophages can be further classified as M2a, M2b, and M2c. During allergy response or parasite infection, IL4 and IL13 induce the production of M2a macrophages, which secrete histamine and to induce encapsulation and destruction of the parasite. Meanwhile, M2b macrophages are activated by immune complexes and TLR/ interleukin 1 receptor (IL1R) ligands. Lastly, M2c macrophages are activated by IL-10, TGF-β, and glucocorticoids (van den Bosch *et al.*, 2017).

Another type of macrophage in humans is the regulatory macrophage (Mreg) (see Figure 2.7).*In vitro*, it was demonstrated that Mregs develop from $CD14^+$ peripheral blood monocytes when exposed to macrophage colony-stimulating factor (M-CSF) and IFNγ for a week. Mregs express molecules such as MHCII, Fc gamma receptor (FcγR), interferon gamma receptor (IFNγR), TLR4, and programmed death ligand 1 (PDL1). They are potent T cell suppressors (van den Bosch *et al.*, 2017).

Dendritic cells (DCs) are called such because of the dendrites or cellular processes extending from their cell body (see Figure 2.10). They are present at low density, but they are widely distributed in the body so that they sense danger signals and mediate immune response or self-tolerance (Germic *et al.*, 2019). Dendritic cells act as messengers between the innate and adaptive immune systems. Similar to neutrophils and macrophages, they have TLRs that detect foreign invaders, which they subsequently ingest. They are also the most potent among the APCs resulting in T and B cell activation. Additionally, they produce pro-inflammatory mediators. Monocyte-derived dendritic cells (MoDCs) are also called inflammatory dendritic cells (iDCs) because they are important in initiating inflammation. They are phenotypically similar to conventional DCs in terms of expression of MHCII, CD11b, and CD11c. However, MoDCs in mice are distinct in their expression of CD64, also known as FCγR1 (De Kleer *et al.*, 2014; Spiering, 2015).

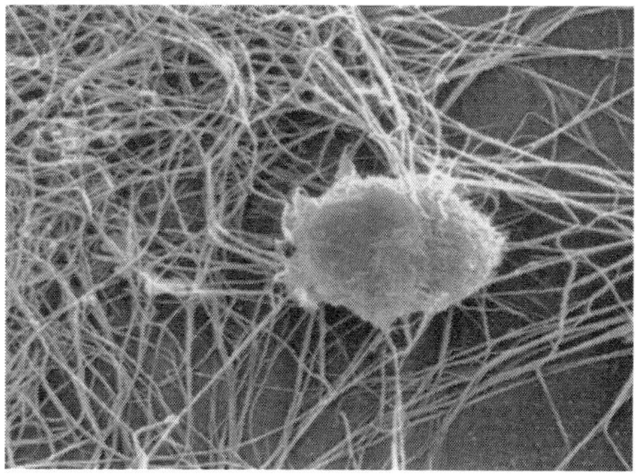

Figure 2.10. A follicular dendritic cell (FDC) has numerous cell processes or dendrites that interconnect with one another and with dendrites of adjacent FDCs. These form a network that can trap antigens and initiate an antibody response. (Source: Askazal, Creative Commons License)

2.1.3.3 Dendritic cells

Apart from monocyte-derived dendritic cells, there are two other two types of dendritic cells (DCs), the classical or conventional (cDCs) and plasmacytoid (pDCs) dendritic cells. It is proposed that these DCs come from hematopoietic precursor cells in the bone marrow (see Figure 2.11). Growth factors such as fms-like tyrosine kinase 3 ligand(Flt3L), granulocyte macrophage colony-stimulating factor (GM-CSF), and M-CSF are needed for the differentiation of the precursor cells to DCs. The earliest precursor is the granulocyte, macrophage, and DC progenitor (GMDP), which has the potential to differentiate into granulocyte, macrophages, and DCs. When this cell initiates the expression of the M-CSF receptor (M-CSFR), it undergoes a change in phenotype, losing the potential to be a granulocyte and becoming a macrophage and DC progenitor (MDP). MDP upregulates CD123 expression to become a common DC progenitor (CDP), which can then differentiate to pDC and cDC depending on the expression of certain transcription factors. The expression of zinc finger and BTB domain continuing 46 (ZBTB46) and inhibitor of DNA binding 2 (ID2) drives the specification into a cDC precursor. Further differentiation into cDC1 depends on basic leucine zipper transcription factor ATF-like 3 (BATF3) and interferon regulatory factor 8 (IRF8), while cDC2 differentiation depends on IRF4 and Krüppel-like factor 4 (KLF4) expression. Meanwhile, transcription factor 4 (TCF4) expression leads to pDC differentiation(Castell-Rodríguez *et al.*, 2017; Rhodes *et al.*, 2019).

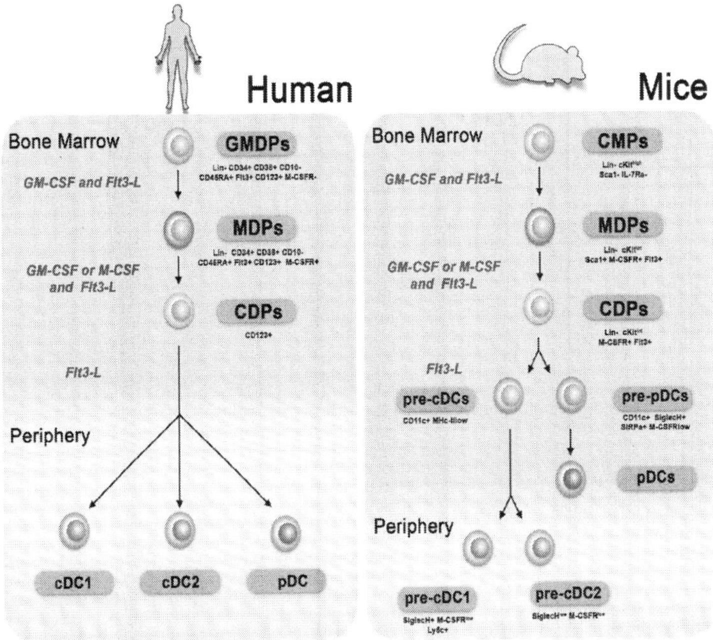

Figure 2.11. There are a series of three progenitor cells, i.e., granulocyte, macrophage, and dendritic cell progenitors (GMDP); macrophage, dendritic cell progenitor (MDP); and common dendritic cell progenitor (CDP) that differentiate in a step-wise manner into classical (cDC) and plasmacytoid dendritic cell (pDC). The differentiation process is influenced by certain growth factors such as fms-like tyrosine kinase 3 ligand (Flt3L), granulocyte macrophage colony-stimulating factor (GM-CSF), and macrophage colony-stimulating factor(M-CSF). (Source: Andrés Castell-Rodríguez *et al*., Creative Commons License)

However, the generation of pDCs from CDPs seems inadequate to explain the frequency of pDCs relative to cDCs. An alternative explanation is that majority of type I IFN-producing pDCs may be generated not from a myeloid progenitor but a pDC-committed lymphoid one. However, other findings reinforce that most pDCs develop from a myeloid lineage and that pDC-biased progenitors arise before their lymphoid counterparts. These results indicate that pDCS can have dual origins, which may dictate their functions when they are fully differentiated. Studies performed in humans and mice also indicate that cDCs can arise from both myeloid and lymphoid precursors. It is proposed that instead of having homogeneous multipotent progenitors that branch out into distinct cell types, progenitors likely have their lineage imprinted during the early stages of development. Therefore, these progenitors follow predetermined differentiation pathways, most following a uni-lineage pathway but some having bi- or multi-lineage potential. Thus, the CDP pool is likely made up of a mixture of progenitors with pre-determined pDC or cDC1/2 fates (Rhodes *et al.*, 2019).

Classical dendritic cells (cDCs) occur in most lymphoid and nonlymphoid tissues (see Figure 2.12). They have an enhanced ability to sense tissue injuries, capture environmental- and cell-associated antigens, and present phagocytosed antigens to T cells. Hence, cDCs induce an immune response to any foreign invaders, but they promote tolerance to self-antigens as well. Their efficiency in performing these functions may be attributed to their location in nonlymphoid tissues and in the spleen marginal zone, which constantly acquire tissue and blood antigens; their superior antigen processing and presentation mechanism; their exceptional ability to migrate while bearing tissue antigens to the T cell zone of lymph nodes both under normal and inflamed conditions; and their superior ability to prime naïve T cells (Merad *et al.*, 2013). There are two families of cDCs, namely the cDC1 and cDC2.

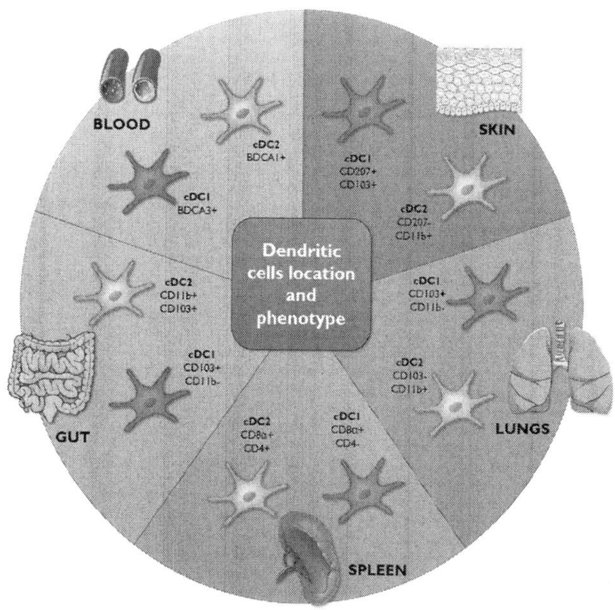

Figure 2.12. The two types of conventional dendritic cells, cDC1 andcDC2, may be located in the blood, skin, mucous membranes of the digestive and respiratory tracts as well as in lymphoid organs such as the spleen. (Source: Andrés Castell-Rodríguez *et al.*, Creative Commons License)

The cDC1 type constitutes approximately 0.05% of peripheral blood mononuclear cells. They are characterized by the expression of CD141/ blood dendritic cell antigen A3$^+$(BDCA3$^+$), C-type lectin-like receptor 9A (Clec9A), and cell adhesion molecule 1 (CADM1). They have a notable cross-presentation ability so that they can efficiently prime CD8T cells against extracellular antigens from bacteria and viruses. Their Clec9A binds to extracellular actin exposed during cellular necrosis such that cDC1s can present necrotic antigens to T cells. Their high levels of TLR3, TLR9, and TLR10 enable them to detect intracellular double-stranded RNA (dsRNA) and DNA and subsequently produce type IIFN(IFNI) and IL12 (De Kleer *et al*., 2014; Rhodes *et al*., 2019).

The cDC2 subset has CD1c/BDCA1$^+$ phenotype. cDC2s make up a great proportion of myeloid DC in the blood. They are potent stimulators of naïve T cells. They express a variety of lectins like Clec4A, Clec10A, Clec12A, and Clec13B (or DEC205). They also express TLR2, TLR4, TLR5, TLR6, TLR8, and TLR9 and when these are stimulated, they produce a wide range of immunoregulatory molecules such as TNFα, IL1, IL6, IL8, IL12, IL18, and chemokines like CCL3, CCL4, and CXCL8. The cDC2 type can be subdivided further into cDC2A and cDC2B. cDC2A exhibits a higher expression of CD11c, CD1c, and MHCII genes. On the other hand, cDC2B has increased expression of inflammatory genes and has a similar phenotype as classical monocytes. It seems that cDC2A undergoes more C-C chemokine receptor 7 (CCR7)-dependent migration, stimulates more naïve T cells to proliferate, and produces greater levels of innate cytokines than cDC2B. cDC2A also appears to induce polarization of Th2, Th17, Th22, and regulatory T (Treg) cell, whereas cDC2B induces Th1 cell polarization (Rhodes *et al*., 2019).

On the other hand, plasmacytoid dendritic cells(pDCs)are so named because of their morphological similarity to plasma B cells in their immature state, i.e. without cytoplasmic projections. They accumulate mainly in the bone marrow, all peripheral organs, and the blood as well as lymphoid tissues and lymph nodes. They are characterized by having the CD123$^+$/ BDCA2$^+$phenotype. They have a highly developed secretory compartment and under normal conditions, they express small amounts of MHCII, costimulatory molecules, and integrin CDC11c. They also express pattern recognition receptors such as TLR7 and TLR9 that recognize foreign single-stranded RNA (ssRNA) and double-stranded DNA (dsDNA), respectively, as well as constitutively express interferon regulatory factor 7 (IRF7). These induce them to produce a great quantity of IFNI and IFNIII, making them highly specialized to respond to viral infections. Activation of their TLR7 and TLR9 also induces nuclear factor of the kappa light chain enhancer of B cells(NF-κB) expression that results in TNFα, IL6, and chemokine (e.g., CCL3, CCL4, CCL5, IL8, CXCL10, CXCL11) production. In their mature state, they are enabled to present antigens and to prime T cells against viral antigens. Otherwise, immature pDCs lack the capacity to stimulate T-cell responses (Merad *et al*., 2013; De Kleer *et al*., 2014; Rhodes *et al*., 2019).

An additional subset of DCs was recently characterized, which is defined by unique identifying markers, namely Axl and sialic acid-binding Ig-like lectin 6 (SIGLEC6). Hence, it is called **Axl$^+$ DC**. This type of DC also exhibits markers characteristic of pDC (i.e., CD123, BDCA2, BDCA4, CD45RA) and cDC2 (CD11c, CD33, CX3CR1, CD1c, CD2). Unlike pDCS, they do not produce IFNI. On the other hand, they resemble the basic function and morphology of cDC2. For instance, they express TLR4, TLR5, IRF4, and IRF8 indicating that they can respond to bacterial infection with cytokine and chemokine production. They are potent T cell stimulators due to their production of high levels of CD86 and human leukocyte antigen-DR isotype (HLA-DR). The SIGLECs they express (i.e., SIGLEC1, SIGLEC2, and SIGLEC6) are glycan-binding lectins, while Axl binds apoptotic cells. This indicates that they may have functions apart from antigen presentation and T cell stimulation (Rhodes *et al*., 2019).

2.2 SIGNALING MOLECULES IN INNATE AND ADAPTIVE IMMUNITY

The function and activity of the cells and other components in the immune system are regulated by certain signaling compounds. These compounds act on other signaling molecules and on cells to trigger other processes, some of which may not be immunological.

Cytokines are small glycoproteins secreted by innate and adaptive immune cells upon contact with a pathogen- or damage-associated molecular pattern or with an antigen. They may also be produced in response to injury. They are produced predominantly by helper T (Th) cells and macrophages. These bind to specific receptors on their own cell surface (autocrine function) or on other immune cells, either located closely (paracrine function) or at a distance (endocrine function). The activation of the cytokine receptors results in either the activation or suppression of the immune cells as well as regulation of their growth and maturation. They are often produced in a cascade, with one cytokine stimulating its target to release additional cytokines. Different cell types can secrete the same cytokine. On the other hand, a single cytokine may act on different cell types (pleiotropic activity). Also, similar functions can be stimulated by different cytokines, and different cytokines can act additively, synergistically, or antagonistically. Cytokines include chemokines, interferons, interleukins, and tumor necrosis factor (Spiering, 2015; Zhang & An, 2007).

Chemokines or chemotactic cytokines promote the movement or chemotaxis of immune cells toward the source or the area with the highest concentration of chemokines (see Figure 2.13). This enables the recruitment of immune cells to sites of inflammation or injury (Spiering, 2015). Therefore, chemokines are essential in controlling infection and wound healing. Chemokine-driven cell migration also enables antigen-specific lymphocytes to enter and survey APCs in lymphoid tissue and ensures immunological tolerance. However, excessive chemokine activation may result in unwanted inflammation that leads to cell death and tissue damage (Gravallese & Monach, 2015; Hughes & Nibbs, 2018). Chemokines also have non-immunological functions, such as hematopoiesis (e.g., the regulation of leukocyte proliferation, survival, and differentiation) as well as organ development. Cancer cells can also express chemokine receptors and respond to chemokines, which can promote their metastasis and angiogenesis (Murphy, 2008; Hughes & Nibbs, 2018). Chemokines are induced by primary pro-inflammatory mediators such as IL1 or TNF (Graves &Jiang, 1995). In turn, the leukocytes recruited by chemokines can also produce other chemokines resulting in the next wave of leukocyte mobilization. Chemokines may also be produced by diseased tissues (Hughes & Nibbs, 2018).

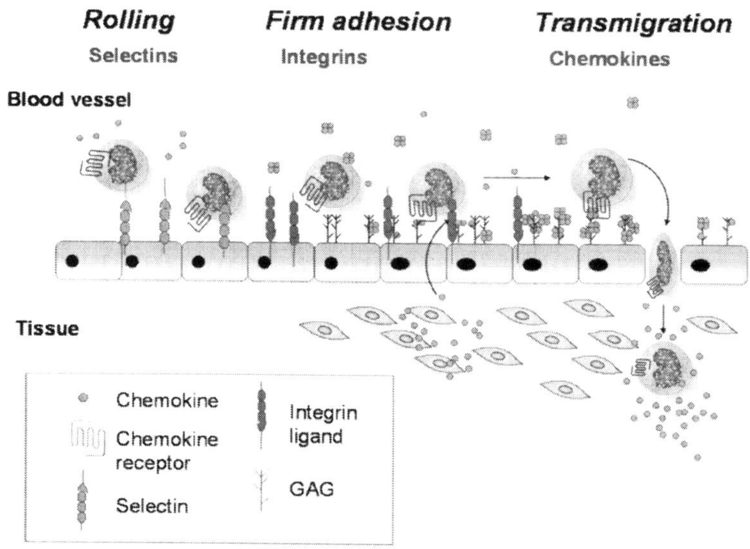

Figure 2.13. Chemokines (yellow circles) are secreted from the blood vessel wall or the underlying tissue in response to inflammatory signals like infection or tissue damage. These are then transported to the luminal surface of the endothelial cells where they bind to their receptors or to chemokine-interacting partners such as glycosaminoglycan (GAG). Thus, the chemokines form an immobilized gradient that drives the migration of leukocytes towards the inflammatory site. (Source: Amanda E.I. Proudfoot*et al.*, Creative Commons License)

The type of leukocyte recruited is largely determined by the nature of the secreted chemokines and corresponding receptors expressed on the migrating cells. There are 50 human chemokines and 23 chemokine receptors on the surface of target cells that have been described, with five of the receptors considered atypical since they do not induce cell migration (Proudfoot *et al.*, 2017). Multiple chemokines, but belonging to the same subclass, may bind to a receptor and, conversely, multiple receptors may bind a given chemokine. Chemokines and their receptors may be assigned to four groups depending on the presence and positions of conserved cysteine (C) residues in their amino (N)-terminal region (Gravallese & Monach, 2015; Murphy, 2008). The C-Cgroup has adjacent cysteines (e.g., regulated upon activation, normal T cell expressed and presumably secreted orRANTES, monocyte chemoattractant protein or MCP-1, macrophage inflammatory protein or MIP-1α, and MIP-1β), C-X-Chas one amino acid between the two N-terminal cysteines (e.g., IL8 or growth related oncogene/

GRO/keratinocyte chemoattractant/KC), Chas one terminal cysteine (e.g., lymphotactin), and CXXXChas three intervening amino acids between the terminal cysteines (e.g., fractalkine) (Zhang & An, 2007; Farooqui, 2018). In addition to chemokine-receptor interaction, chemokines also interact with glycosaminoglycan (GAG) chains of endothelial proteoglycans (see Figure 2.13). The binding of chemokines to GAGs creates an immobilized gradient that gives directional signals for the leukocyte migration towards the inflammatory site. It also restricts the encounter between a chemokine and its receptor at a surface near the chemokine source, and this is thought to prevent the leukocytes from being activated before they reach the inflammatory site (Proudfoot *et al.*, 2017).

Interferons (IFNs) are named such because of their ability to interfere with viral replication (Ferreira *et al.*, 2018). They are released by cells in response to Toll-like receptor (TLR) activation by microbial products, nucleic acids, lipids, polysaccharides, or proteins. They bind the receptors of neighboring cells, consequently activating them. Such activation results in increased expression of more than 2, 000 interferon-stimulated genes (ISGs) to produce more proteins that enable the cells to detect and neutralize pathogens (see Figure 2.14). Examples of these are 2', 5'-oligoadenylate synthetases (OASs) and ribonuclease L (RNASEL), which inhibit a broad range of viruses. IFNs can also promote the mobilization of leukocytes to infected sites by promoting the expression of vascular adhesion molecules and inducing the production of chemokines. They can also stimulate ISGs that regulate normal survival and death of normal and tumor cells, such as p56-related proteins. IFNs are also involved in the activation of inflammasomes, which are cytosolic multiprotein complexes made up of a sensor, apoptosis-associated speck-like protein containing a CARD domain (ASC) that serves as an adaptor protein, pro-caspase 1, and a member of the NOD-like receptor (NLR) family. The activated pro-caspase 1 cleaves pro-inflammatory cytokines, leading to their maturation and release. Inflammasomes also initiate pyroptic or inflammation-associated cell death (Borden *et al.*, 2007; Kopitar-Jerala, 2017; Borden *et al.*, 2008).

There are three types of interferons based on their target receptors, namely type I (IFNI), II (IFNII), and III (IFNIII). There are 17IFNIs in humans including the IFNα family with 13 subtypes, IFNβ, IFNω, IFNT, IFNκ, and IFNε.In general, IFNIs are predominantly produced by pDCs and also by T cells, monocytes, fibroblasts, and epithelial cells. Specifically, leukocytes produce the αsubtype while fibroblasts and DCs produce IFNβ. Under healthy conditions, IFNIs are present at undetectable levels. Their synthesis can be triggered by virus infection (see Figure 2.14) or viral gene products, like dsRNA, ssRNA, dsDNA or viral envelope proteins, which are detected by TLRs, retinoic acid-inducible gene I-like receptors (RLRs), and cyclic GMP-AMP synthase (cGAS). It can also be induced by bacterial genetic materials. IFNII includes IFNγ, which is produced by lymphocytes, T cells, and macrophages. It is structurally different from type I and uses a different receptor. The most recently discovered IFNIIIs are also called IFNλor IL28/29. After virus infection, it is mainly pDCs that produce this type of IFN. They are similar to type I in terms of structure and induction characteristics, but they function through a different receptor (Spiering, 2015; Ferreira *et al.*, 2018; Borden*et al.*, 2008; Rehwinkel &Gack, 2020).

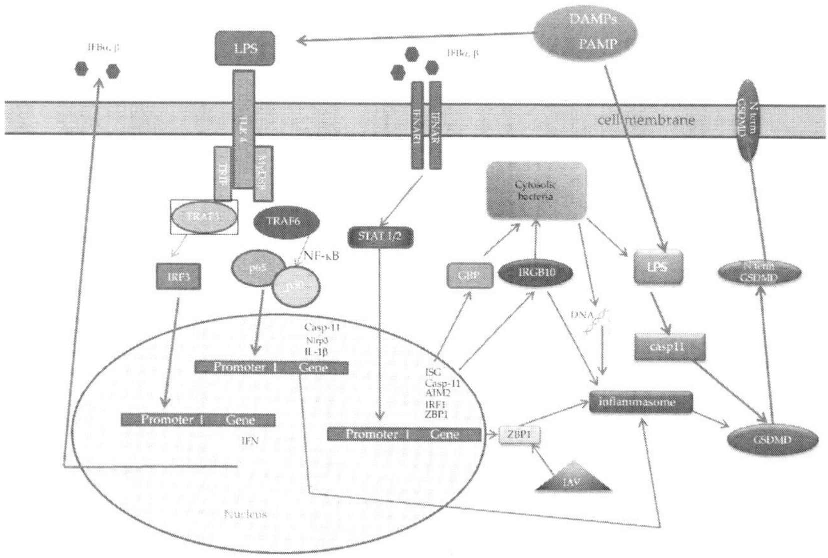

Figure 2.14. Interferon (IFN) signaling influences the recognition of intracellular pathogens, such as bacteria and influenza A virus (IAV). Type I IFN activates the transcription factor interferon regulatory factor 1 (IRF1), which initiates the expression of guanylate-binding proteins (GBPs) and interferon response gene B10 (IRGB10). These gene products permeabilize the membrane of Gram-negative bacteria resulting in the release of bacterial DNA and lipopolysaccharide (LPS). AIM2 inflammasome detects bacterial DNA while caspase 11 (casp11) directly interacts with LPS. Type I IFN signaling also mediates the increased expression of Z-DNA-binding protein 1 (ZBP1), which recognizes the IAV proteins and triggers inflammasome activation and induces apoptosis, necrosis, and pyroptosis of IAV-infected cells. (Source: Nataša Kopitar-Jerala, Creative Commons License)

IFNα and β and the other IFNIsshare a common receptor, the interferon α/β receptor (IFNAR). The bound IFNAR activates the Janus kinase-signal transducer and activator of transcription (JAK-STAT) signaling pathway. In particular, STAT1 and STAT2 are phosphorylated inducing them to interact with interferon regulatory factor 9 (IRF9) to form the interferon-stimulated gene factor 3 (ISGF3) transcriptional complex. ISFG3 binds to type I interferon-dependent gene promoters and initiates the transcription of type I ISGs (Rehwinkel & Gack, 2020). Both IFNα and β stimulate NK cells to increase their lysis potential. In addition, IFNIs significantly upregulate MHCI-dependent antigen presentation and stimulate Th1 cell development. IFNα can also be used to treat certain types of cancers. It enhances the ability of certain cells to attack cancer cells as well as slows down the growth of cancer cells and the development of blood vessels in tumors (American Cancer Society, 2019; Borden *et* al., 2007; Muñoz-Carrillo *et al*., 2018; Dafny *et al*., 2004).

IFNγ, on the other hand, promotes the differentiation of B cells into plasma cells producing IgG. It also induces phagocytosis by activating macrophages, increases the expression of MHCI and MHC II on APCs, promotes complement activation, increases the cytotoxic activity of T cells, and enhances Th1 cell differentiation (Muñoz-Carrillo *et al*., 2018).IFNIIIs, just like type I, are essential in inhibiting viral replication (Ferreira *et al*., 2018).

At low concentrations, interleukins (IL) are proteins that facilitate communication among leukocytes under inflammatory conditions. They could either be pro- or anti-inflammatory. At higher concentrations, some ILs enter the bloodstream and act as endocrine hormones, resulting in fever and stimulating the production of immune proteins in the liver. These are produced by leukocytes, lymphocytes, monocytes, macrophages, and, sometimes, even non-immune cells. ILs are grouped into families based on sequence homology and receptor chain similarities or functional properties (Spiering, 2015; Ferreira *et al*., 2018). The IL1 superfamily of ligands and receptors is associated with acute and chronic inflammation. The IL1 cytokines increase nonspecific resistance to infection. The IL6 family of cytokines (i.e., IL6, IL27, IL31, IL35) mainly induces the production of inflammation proteins during the acute phase of inflammation, differentiation of B cell into a plasma cell, Tcell modulation, Th17 cell development, and hematopoiesis. Another family of cytokines is the IL17, which is mainly pro-inflammatory but may exert anti-inflammatory effects in certain conditions. These cytokines are important in mounting a defense against extracellular pathogens. The type I cytokine superfamily, which includes the IL2 family, is also known as hematopoietins (i.e., IL2, IL3, IL4, IL6, IL7, IL9, IL12, IL15, IL21)and mainly functions for the growth and differentiation of precursor leukocytes into mature ones apart from the initiation and modulation of inflammation. IL10 is a potent pro-inflammatory cytokine that performs its function through the activation of STAT1, STAT3, PI3K, and P38 mitogen-activated protein kinases (MAPK) pathways. It is critical for the suppression of Th1 cytokines, and it is also capable of antigen presentation (Muñoz-Carrillo *et al.*, 2018; Ferreira *et al.*, 2018).

Tumor necrosis factor (TNF) derives its name from its ability to cause necrosis of tumors *in vitro*. Its expression is stimulated by the binding of lipopolysaccharide and other pathogenic material to TLRs and also by other cytokines, such as IFNγ. TNFα, also known as cachectin, is a major pro-inflammatory cytokine, promoting inflammation during infection and in dysregulated immune responses, such as in arthritis or graft rejection. It is primarily produced by macrophages during the acute phase of inflammation. It is also produced by other immune cells including T cells, NK cells, and monocytes. It binds to its specific receptor, TNFR or CD120, and then activates many transcription factors like NF-κB, which enhances the expression of pro-inflammatory genes. It also stimulates chemokine production, induces apoptosis and necrosis in some cell types or certain forms of tumors as well as activates stress-activated protein kinases (SAPKs) (Spiering, 2015; Ferreira *et al.*, 2018; Zhang& An, 2007; Muñoz-Carillo *et al.*, 2018). Because TNFα activates several signaling pathways, it regulates different biological processes, namely stimulation of the immune system, resistance to pathogens, resistance to tumors, sleep regulation, and embryonic development. However, parasites, bacteria, and viral infections become more pathogenic and fatal due to TNF circulation. Also, TNF along with interferon exhibit cytotoxicity even to nonmalignant cells so that high concentrations of TNFα are toxic (Muñoz-Carrillo *et al.*, 2018).

TNFα has also been shown to mediate anti-inflammatory and immunomodulatory effects through several possible mechanisms(see Figure 2.15). Chronic stimulation with TNFα has negative effects on conventional T cells. It may inactivate T cell receptor (TCR) signaling, induce T cell exhaustion, or kill CD8 T cells. TNFα may also boost cells with immunosuppressive activity, such as myeloid-derived suppressor cells (MDSCs), by promoting their survival, local recruitment, or suppressive function. Lastly, the immunosuppressive effect of TNFα may be related to its stimulation of Treg proliferation, survival, and stability (Salomon *et al.*, 2018).

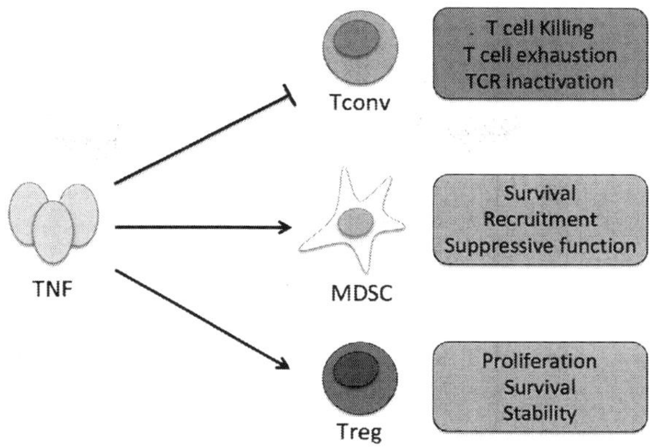

Figure 2.15. Tumor necrosis factor α (TNF) exhibits immunosuppressive function via its intrinsic negative effect on conventional T cells (Tconvs) or its stimulatory effect on suppressive cells, such as the myeloid-derived suppressor cell (MDSC) or regulatory T cell (Treg). (Source: Beno I t L. Salomon, Creative Commons License)

Apart from TNFα and its receptor, the tumor necrosis factor superfamily has 18 other ligands and 28 other receptors. Members of this family are essential in immunity, inflammation, cell cycle regulation, cell proliferation, cell differentiation, and apoptosis (Muñoz-Carrillo *et al.*, 2018).

2.3 PRR-PAMP/DAMP MECHANISM OF INNATE IMMUNE RESPONSES

The innate immune system is relatively nonspecific. It relies on pattern recognition receptors (PRRs), which enable a limited range of immune cells to recognize and respond quickly to a wide range of pathogens that share common and conserved structures called pathogen-associated molecular patterns (PAMPs) or to molecules released by damaged or aberrant host cells known as damage-associated molecular patterns (DAMPs) (see Figure 2.16). Examples of PAMPs include carbohydrates, proteins, or nucleic acids that are specific to the pathogens and not present in the host. DAMPs, on the other hand, are signals produced in the absence of microbial components. They arise from the necrosis of healthy cells, and sometimes from apoptotic cells, in response to inflammation, ischemia, or hypoxia. They may also be produced extracellularly by leukocytes and other cells in response to microbial components, cytokines, or cellular stress (Marshall *et al.*, 2018; Kato & Svennson, 2015; Amarante-Mendes *et al.*, 2018; Sue *et al.*, 2018).

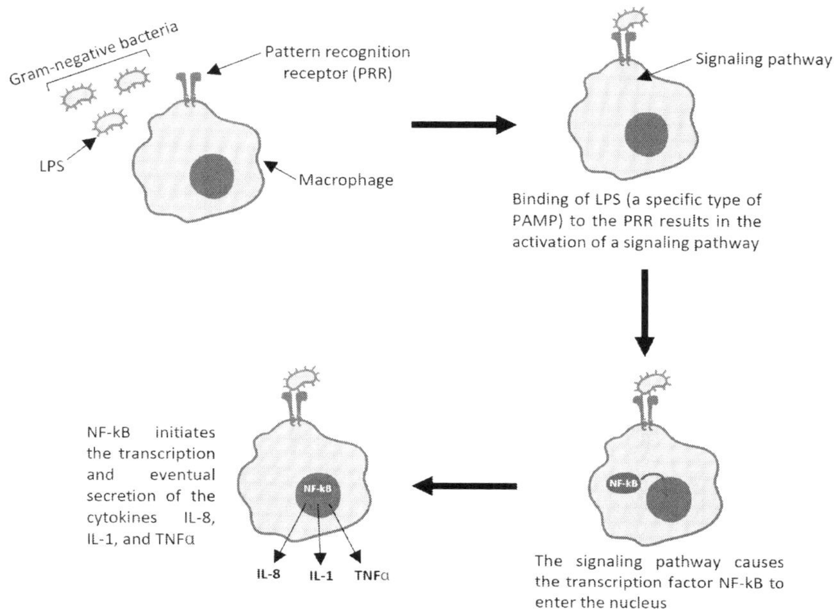

Figure 2.16. A pattern recognition receptor (PRR) recognizes and binds certain structures common to pathogens such as lipopolysaccharide (LPS) in the membrane of Gram-negative bacteria. This activates a signaling pathway that upregulates the expression of cytokine genes, eventually resulting in the secretion of cytokines. (Source: Immcarle105, Creative Commons License)

The engagement of PAMPs or DAMPs to PRRs induces signals that activate a multitude of intracellular signaling pathways, including adaptor molecules, kinases, and transcription factors (see Figure 2.16). These promote gene expression and synthesis of a broad range of molecules like cytokines, chemokines, cell adhesion molecules, and immunoreceptors. As a result, adaptive immune cells as well as anti-microbial and pro-inflammatory responses are activated so that the infectious agents are eliminated or at least, contained (Sue *et al.*, 2018; Mogensen, 2009). PRRs can be found on immune and also non-immune cells, particularly on the plasma membrane to recognize extracellular pathogens and on endosomal membranes as well as the cytosol to detect intracellular invaders. They can also occur extracellularly when secreted in the blood and interstitial fluids (Amarante-Mendes *et al.*, 2018). Different pathogens, i.e., viruses, bacteria, fungi, and protozoa, are recognized by different PRRs, yet the mechanisms involved are similar and overlapping (Mogensen, 2009).

2.3.1. The Different Families of Pattern Recognition Receptors

There are four major families of PRRs, namely the Toll-like receptors, the nucleotide-binding oligomerization domain-leucin rich repeats-containing receptors (NLRs), the retinoic acid-inducible gene I-like receptors (RLRs) also known as RIG-I-like helicases (RLHs), and the C-type lectin receptors (CLRs). These four classes of PRRs differ in their ligand recognition, signal transduction, and sub-cellular localization. They can also function coordinately ensuring optimum response (Amarante-Mendes *et al.*, 2018; Zaru, 2021).

2.3.1.1. Toll-like receptors

The most extensively studied PRRs are the Toll-like receptors (TLRs). Their name was derived from their homology to the *Drosophila melanogaster* Toll protein, which is a receptor involved in the antifungal response. These are transmembrane proteins located in the plasma or endosomal membranes, enabling them to detect PAMPs in the extracellular environment or those that reached the endosomes through endocytosis, respectively (Zaru, 2021; Takeda & Akira, 2005). They serve to link the innate and adaptive immune systems. These are present on respiratory and gastrointestinal epithelial cells, APCs, hematopoietic stem cells, mast cells, Tregs, NK cells, and endothelial cells. A TLR is made up of a leucine-rich repeat (LRR) extracellular domain that mediates the recognition of PAMPs, a transmembrane domain, and a cytoplasmic Toll/IL-1 receptor (TIR) domain that initiates downstream signaling (see Figure 2.17). TLRs are in the form of a homo- or heterodimer as they interact with their respective PAMPs or DAMPs, and they are aided by a co-receptor or accessory molecule (Kawasaki & Kawai, 2014).

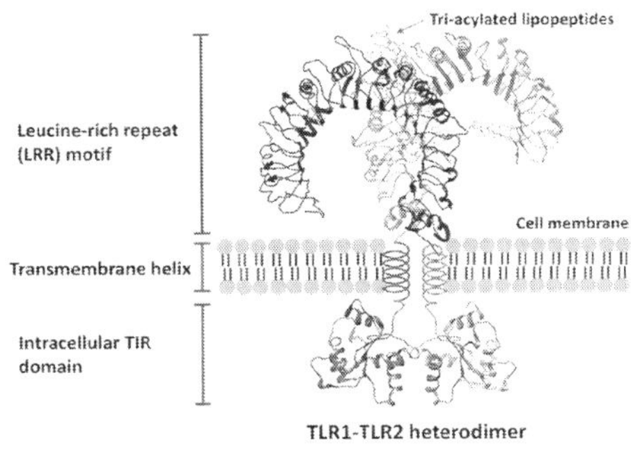

Figure 2.17. A typical Toll-like receptor (TLR) has three important regions, namely the horseshoe-shaped leucine-rich repeat (LRR) motif that protrudes

into the extracellular side, the transmembrane helix, and the intracellular Toll/IL-1 receptor (TIR) domain. TLRs dimerize when activated forming homodimers or in this case, a heterodimer (i.e., between TLR1 and TLR2). (Source: Wei Gao *et al*., Creative Commons License)

There are 10 human TLRs that have been identified. TLR1, TLR2, TLR4, TLR5, TLR6, and TLR10 are located in the plasma membrane, while TLR7, TLR8 and TLR9 are found in the endosomal membrane (Sue *et al*., 2018) (see Figure 2.18). The endosomal TLRs have their LRR domain facing the luminal side. TLR3 may be found on the cell surface and in the membrane of endosomes (Mogensen, 2009). Cell surface TLRs mainly recognize membrane components of microbes, such as lipids, proteins, and lipoproteins. Meanwhile, intracellular TLRs sense bacterial and viral nucleic acids as well as self-nucleic acids in disease conditions like autoimmunity (Kawasaki & Kawai, 2014).

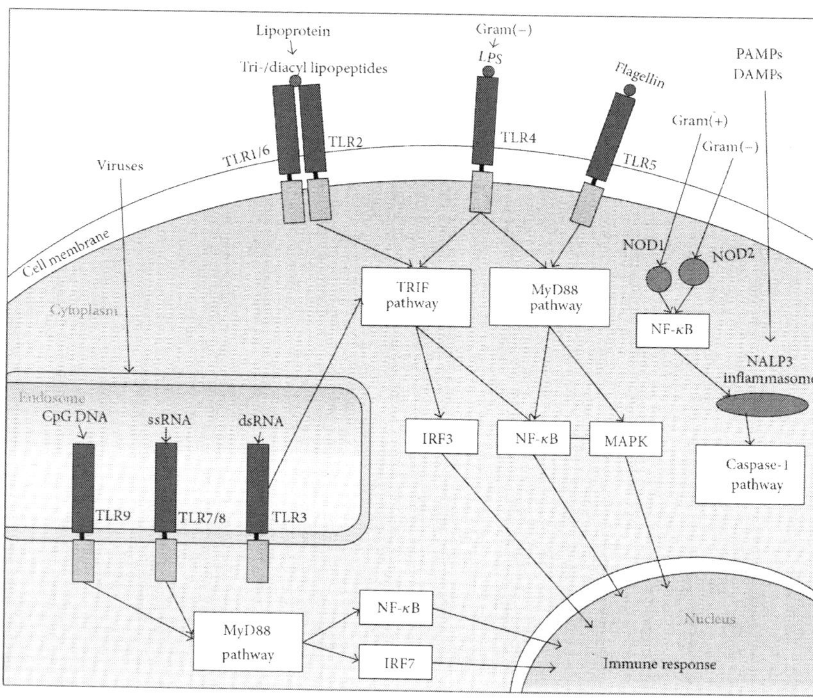

Figure 2.18. Toll-like receptors (TLRs) are transmembrane receptors while nucleotide-binding oligomerization domain-leucin rich repeats-containing receptors (NLRs) are cytosolic ones. The different types of TLRs and NLRs (i.e., NOD1, NOD2, and NALP3 inflammasome) recognize certain pathogen associated molecular patterns (PAMPs) and damage-associated molecular patterns

(DAMPs), activating signaling pathways. The TLR1/TLR2 and TLR6/TLR2 heterodimers, TLR4, and TLR3 trigger the TIR-domain-containing adapter-inducing interferon-β (TRIF) pathway causing the activation of interferon regulatory factor 3 (IRF3) andNF-κB. TLR5, TLR9, and TLR7 stimulate the myeloid differentiation primary-response gene 88(MyD88) pathway activating NF-κB, IRF7, and mitogen-activated protein kinase (MAPK). NOD1 and NO-D2activate NF-κB, while NACHT, LRR, PYD domains-containing protein 3 (NALP3) inflammasome recruits and activates the caspase-1 pathway. (Source: Ji-Hyun Jang *et al.*, Creative Commons License)

TLR2recognizes the peptidoglycans and lipoproteins of Gram-positive as well as certain carbohydrate and protein components of viruses, bacteria, and fungi (see Figure 2.18). On the other hand, TLR3 interacts with the dsRNA of viruses, small interfering RNAs, and self-RNA from damaged cells. TLR4 detects lipopolysaccharide of Gram-negative bacteria. It also interacts with envelope glycoproteins of viruses, glycoinositolphospholipids of protozoa, and mannan of *Candida*. It can also be activated by hsp60, hsp70, and fibronectin that are present in sites of inflammation. Meanwhile, TLR5 recognizes flagellin, whereas TLR7 and TLR8 detect ssRNA of RNA viruses and certain bacteria. TLR 9 identifies CpG motifs within the viral and bacterial genomes. Finally, TLR 10 detects influenza virus A. TLR2 may associate with another type of TLR to form heterodimers. For instance, TLR1/TLR2 detects bacterial triacyl lipoproteins and TLR2/TLR6 detects diacyl lipoproteins of *Mycoplasma*, whereas TLR2/TLR10 recognizes PAMPs from listeria(Sue *et al.*, 2018; Mogensen, 2009; Kawasaki & Kawai, 2014). In the case of DAMPs, they are recognized by TLR2, TLR4, and TLR7 (Kato & Svennson, 2015).

TLRs can either interact with PAMPs directly or through an intermediate PAMP-binding molecule. As an example, TLR1/TLR2, TLR3, and TLR9 directly bind to their respective PAMPs directly, but TLR4 does so through the assistance of the myeloid differentiation factor 2 (MD2) molecule. A given pathogen can have different PAMPs so that it can activate several types of TLRs. Likewise, structurally unrelated pathogens can activate a given TLR(Mogensen, 2009).

The binding of PAMPs or DAMPs to TLR induces TLR oligomerization, which subsequently activates downstream signaling pathways(see Figure 2.18) (Mogensen, 2009). Specifically, the adaptor protein myeloid differentiation factor 88 (MyD88)binds to the TIR domain of TLRs and recruits the serine/threonine kinase IRAK4 to TLRs. Then IRAK4 phosphorylates IRAK1 leading to the activation of the latter. The activated

IRAK1 associates with TNFR-associated factor 6 (TRAF6), which induces two signaling pathways. One of these involves MAPK-mediated activation of the activator protein 1 (AP1) transcription factors (Takeda & Akira, 2005). Another pathway activates the TAK1/TAB complex, a member of the MAPK family. This complex stimulates the IκB kinase (IKK) complex to phosphorylate the NF-κBinhibitor protein (IκB) that are bound to NF-κB in the cytoplasm. The phosphorylation of IκB triggers their subsequent degradation so that NF-κB dimers are released and transported into the nucleus (Takeda & Akira, 2005; Torres *et al*., 2011). The activation and nuclear translocation of AP1 and NF-κB result in the expression of genes for pro-inflammatory cytokines, such as TNFα, IL1, IL6, and certain chemokines. Bound TLRs also recruit and trigger Toll-IL-1 receptor domain containing adaptor-inducing interferon β (TRIF). TRIF signaling pathway also activates NF-κB as well as the transcription factor IFN regulatory factor 3 (IRF3) leading to the transcription of type I IFN genes. This ultimately results in the eradication of viruses (Sue *et al*., 2018; Kawasaki & Kawai, 2014).

TLRs have also been linked to phagocytosis. Studies showed that without TLR2/TLR4 or MyD88, there is impaired phagosome maturation after bacterial phagocytosis. Furthermore, it has been observed that p38 activation by TLR-mediated MyD88 signaling pathway is necessary for phagosome maturation (Takeda & Akira, 2005).

2.3.1.2. Nucleotide-binding oligomerization domain-leucin rich repeats-containing receptors

Nucleotide-binding oligomerization domain-leucin rich repeats-containing receptors (NLRs) are cytoplasmic sensors characterized by a centrally located nucleotide binding oligomerization domain (NOD) that induces oligomerization, recruitment of downstream signaling molecules, and increased expression of inflammatory genes (see Figure 2.19). They also have an LRR C-terminal region that mediates ligand sensing and an N-terminal caspase activation and recruitment domain (CARD) for signal initiation (Muñoz-Carrillo, 2017; Mogensen, 2009).

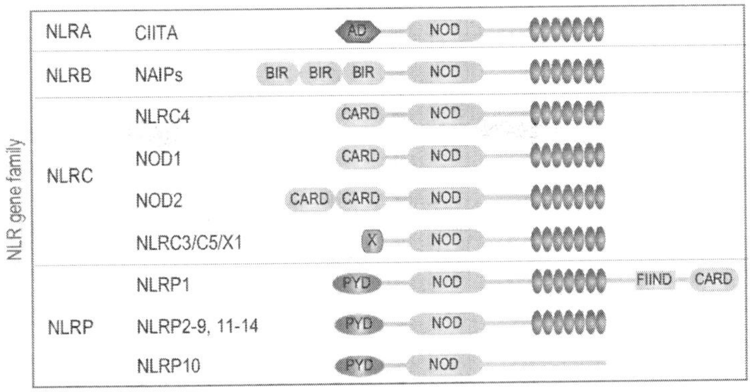

Figure 2.19. Nucleotide-binding oligomerization domain leucine-rich repeats containing receptors (NLRs) have common domains, namely the nucleotide-binding oligomerization domain (NOD) at the middle portion and the leucine-rich repeats (LRR) domain at the C-terminal region (green). Human NLRs are classified based on their domains at the N-terminus. NLRA has an acidic trans-activation domain (AD), NLRB has a baculovirus inhibitor of apoptosis protein repeat (BIR), NLRC has a caspase-recruitment and activation domain (CARD), and NLRP has a pyrin domain (PYD). (Source: Yifei Zhong *et al.*, Creative Commons License)

There are 23 NLRs in humans that belong to different subfamilies, which are usually identified by the signaling domains at their N-terminal regions (see Figure 2.19). NLRA subfamily contains an acidic transactivation domain (AD), NLRB has a baculovirus inhibitor of apoptosis protein repeat (BIR) domain, NLRC has a CARD domain, NLRP contains a pyrin domain (PD), and NLRX is named such because it has an unknown domain (Takeda & Akira, 2005). When the NLR is bound by PAMPs, the auto-inhibitory LRR changes in conformation so that the N-terminal domain is exposed and allowed to interact with downstream signaling adaptors or effectors. This enables NLR oligomerization forming large molecular scaffolds (Zhong *et al.*, 2013).

NLRC includes NOD1 (or NLRC1) with one CARD domain and NOD2 (or NLRC2) with a couple of CARD domains. These are the best-described members of the NLR family. They detect molecules from the synthesis and degradation of bacterial peptidoglycan (see Figure 2.20). Specifically, NOD1 recognizes γ-D-glutamyl-meso diaminopimelic acid (iE-DAP) that occurs in Gram-negative bacteria and in some Gram-positive ones, while NOD2 recognizes the muramyl dipeptide (MDP), MurNAc-L-Ala-D-isoGln, that is found in almost all bacteria. The binding of these peptidoglycan structures to NOD1 and NOD2 induces these receptors to self-oligomerize. In turn, the receptor-interacting serine/threonine protein kinase 2 (Rip2 or RICK or RIPK2) is activated.RIPK2 has its own CARD domain that interacts with the NOD CARD domain. Such association recruits E3 ubiquitin ligases like cellular inhibitor of apoptosis 1(cIAP1), cIAP2, X-linked inhibitor of apoptosis (XIAP), and ITCH resulting in the ubiquitination of RIPK2. This leads to the activation and recruitment of transforming growth factor beta-activated kinase 1(TAK1) complex to the central region of RIPK2 resulting in the activating phosphorylation of NF-κB, AP1, and MAPK. This will then induce the expression of inflammatory cytokines and antimicrobial genes. The ubiquitinated RIPK2 also interacts with another E3 ligase complex called linear ubiquitin assembly complex (LUBAC). LUBAC attaches linear ubiquitin chains to the regulatory protein NF-κB essential modulator(NEMO), also known as IKKα/β/γ, which then activates the IKK complex. The kinase activity of IKK activates the NF-κB signaling pathway (Takeda & Akira, 2005; Muñoz-Carrillo, 2017; Mogensen, 2009; Zhong *et al.*, 2013).

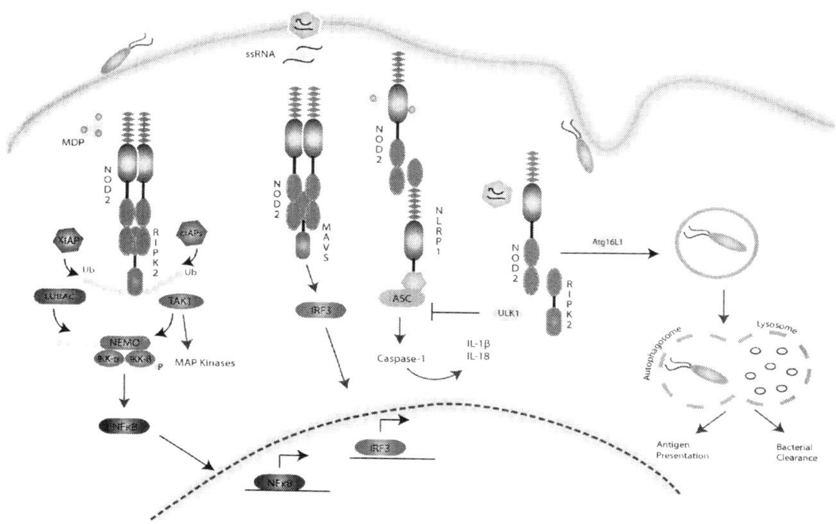

Figure 2.20. The binding of NOD2 to the bacterial peptidoglycan derivative muramyl dipeptide (MDP) induces it to form an oligomer with another NOD2. The oligomer interacts with receptor-interacting serine/threonine protein kinase 2 (RIPK2) resulting in the recruitment of E3 ubiquitin ligases like cellular inhibitor of apoptosis 1 (cIAP1), cIAP2, and X-linked inhibitor of apoptosis (XIAP). The ubiquitinated RIPK2 serves to engage and activate TAK1 and IKK to induce the NF-κB and MAPK signaling pathways. The NOD receptors are

also thought to be activated by single-stranded RNA (ssRNA) viruses so that they interact with mitochondrial antiviral-signaling protein (MAVS) and activate the IRF3 pathway. They can also interact with NLRP via their CARD domains to activate the inflammasome. (Source: Yifei Zhong *et al.*, Creative Commons License)

NOD2 is possibly activated by ssRNA viruses as well (see Figure 2.20). It interacts with mitochondrial antiviral-signaling (MAVS) protein so that IRF3 signaling is activated. This results to the production of IFNIs (Zhong *et al.*, 2003).

NLR also forms part of the inflammasome complex. Apart from NLR, an adaptor ASC and pro-caspase 1 also make up an inflammasome (Zaru, 2021; Muñoz-Carrillo, 2017; Mogensen, 2009) (see Figure 2.21). CARD-CARD interaction mediates oligomerization of the NLR components (see Figure 2.20). This causes proteolysis ofpro-caspase 1 to form the active caspase, which, in turn, cleaves IL1 precursors eventually producing biologically active IL1. There are several families of inflammasomes, each recognizing certain PAMPs through their respective NLR (Mogensen, 2009). The three main types are Nacht LRR protein 3 (NALP3 or NLRP3), NALP1 or NLRP1, and ice protease-activating factor (IPAF) or NLRC4 inflammasome (Takeda & Akira, 2005). The CARD domain of NLRC4 enables direct interaction with caspase 1 even without ASC (see Figure 2.21). NAIPs, which are members of the NLRB subfamily, are important components of the NLRC4 inflammasome (Zhong *et al.*, 2003).

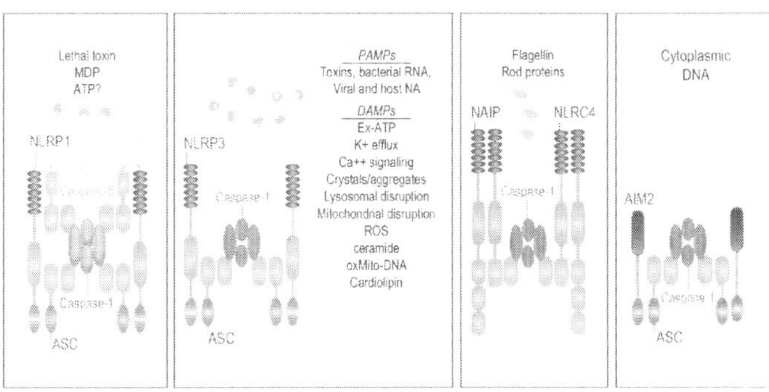

Figure 2.21. There are three main classes of NLR inflammasomes. The NLRP1 inflammasome is made up of NLRP1, the adaptor apoptosis-associated speck-like protein containing a CARD domain (ASC),caspase-1, and caspase-5. It is possibly activated by the anthrax lethal toxin, bacterial muramyl dipeptide (MDP), and decreased cytosolic ATP. NLRP3 is stimulated by a wide variety of

PAMPs and DAMPs. NLRC4 inflammasome responds to bacterial flagellin and type III secretion system (T3SS) proteins. (Source: Yifei Zhong *et al.*, Creative Commons License)

The best-characterized inflammasome is the NACHT, LRR, PYD domains-containing protein 3 (NLRP3), which is essential in inflammation and antiviral responses. The mitochondrial protein MAVS interacts with NLRP3 so that it is recruited to the mitochondrion. This promotes NLRP3 assembly and activation (Wang *et al.*, 2017; Zhao & Zhao, 2020). There are two steps involved in the activation of NRLP3 (see Figure 2.22). The first one is the priming step induced by PRR, TNFR, or IFN receptor (IFNR) activation as well as by pro-inflammatory cytokines. This, in turn, activates NF-κB and promotes the expression of NLRP3, pro-IL-1β, and pro-IL-18. Priming is followed by the activation step, which is triggered by a variety of DAMPs and PAMPs produced during infections, tissue damage, or metabolic imbalances. These include pore-forming toxins, nucleic acids, lipopolysaccharide, peptidoglycan, lipoteichoic acid, and invading pathogens. In this step, NLRP3 recruits ASC through its N-terminal pyrin domain (PYD) resulting in the formation of ASC oligomers (Zhao & Zhao, 2020).

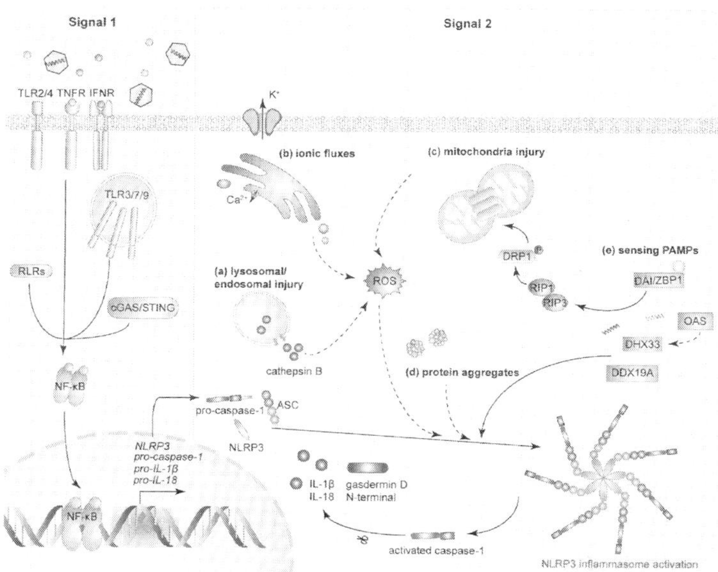

Figure 2.22. During viral infections, the NLRP3 inflammasome is activated through a two-signal mechanism. The priming signal is generated by the ac-

tivation of pattern recognition receptors (PRRs), tumor necrosis factor receptor (TNFR), or interferon receptor (IFNR). This promotes the transcription and synthesis of NLRP3, pro-caspase 1, pro-IL1β, and pro-IL18. The second signal or the activation signal is initiated by DAMPs and PAMPs that induce the assembly and activation of the inflammasome. DAMPs include (a) lysosomal and endosomal injury, (b) aberrant ionic influxes, (c) mitochondrial injury, and (d) protein aggregates. (e) PAMPs, such as viral proteins and RNA, are detected by DAI/ZBP1, DHX33, OAS, or DDX19A leading to the activation of NLRP3 inflammasome. This results in the autocleavage of pro-caspase 1 and the resulting caspase 1 mediates the cleavage of pro-IL1β, pro-IL18, and gasdermin D (GSDMD) into their active forms. (Source: Chunyuan Zhao & Wei Zhao, 2020)

The assembly and activation of NLRP3trigger the autocleavage of pro-caspase 1 (see Figure 2.22). This produces the active caspase 1. Caspase activation is enhanced by ATP, which is released by pathogens, damaged cells, or monocytes. Caspase catalyzes the proteolytic processing of pro-IL1β, pro-IL18, and the propyroptic factor gasdermin D (GSDMD) into their activated counterparts (Takeda & Akira, 2005; Zhao & Zhao, 2020). GSDMD forms pores in the membrane of infected cells so that IL1β and IL18 can be secreted and pyroptosis that is a very rapid inflammatory cell death is induced. The secreted IL-1β subsequently recruits neutrophils to the site of inflammation to help eliminate the viruses. In addition, IL1β and IL18 induce adaptive immune responses (Zhao & Zhao, 2020).

NALP1 inflammasome and NOD2arenecessary for the release of IL1β in response to *Bacillus anthracis* lethal toxin (see Figure 2.21). NALP1 is also triggered by MDP and reduced levels of cytosolic ATP. IPAF inflammasome, on the other hand, is activated by bacterial flagellin and the rod and needle components of the bacterial type III secretion system (T3SS). The resulting activation of caspase 1 leads to the release of IL1β (Takeda & Akira, 2005; Zhong *et al.*, 2003).

2.3.1.3. Retinoic acid-inducible gene I-like receptors

Retinoic acid-inducible gene I-like receptors(RLRs) are located in the cytoplasm and recognize viral RNA from actively replicating viruses. When activated, they trigger NF-κB, MAPK, and IRF intracellular signaling resulting in antiviral responses, such as the production of IFNI and inflammatory cytokines. The activation of these signaling pathways is mediated by adapter molecule mitochondrial antiviral-signaling (MAVS) protein. MAVS is known by three other names, like IFNβ promoter stimulator 1 (IPS1), CARD adapter inducing IFNβ (CARDIF), or virus-induced signaling adapter (VISA). MAVS contains a CARD-like domain in its N-terminal region that interacts with the RLRCARD domain (see Figure 2.23E) and a transmembrane domain at the C-terminus that targets the protein to the mitochondrion and triggers antiviral responses (Zaru, 2021; Takeda & Akira, 2005).

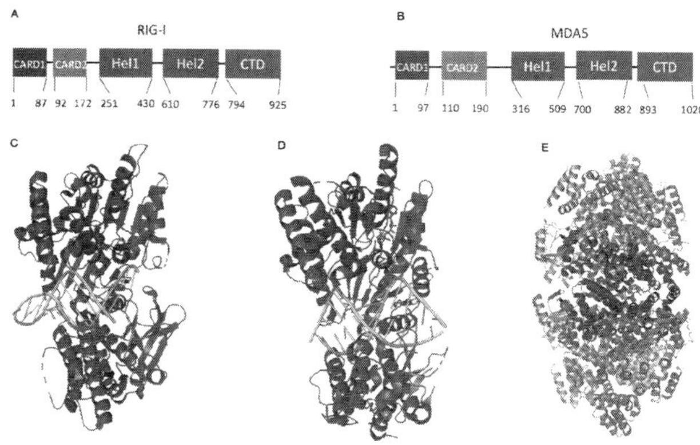

Figure 2.23. (A) retinoic acid-inducible gene I (RIG-I) and (B) melanoma differentiation-associated gene 5 (MDA5) both have two CARD domains at their N-terminal region , two centrally located helicase domains (Hel1 and Hel2), and a carboxy-terminal domain (CTD). The CTD (purple)binds the viral dsRNA (light green) in both (C) RIG-I and (D) MDA5. (E) The RLR CARD domains (blue and aquamarine) interact with the CARD-like domain (orange) of the adaptor molecule mitochondrial anti-viral signaling (MAVS) protein. (Source: Morgan Brisse & Hinh Ly, Creative Commons License)

There are three types of RLRs, namely retinoic acid-inducible gene I (RIG-I), melanoma differentiation-associated gene 5 (MDA5), and laboratory of genetics and physiology 2 (LGP2). They are highly helical having a central DExD/H box RNA helicase with two RecA-like helicase domains (Hel1 and Hel2) and ATP-binding motif and a carboxy-terminal domain (CTD) that binds to viral RNA ligands (see Figure 2.23A-D). RIG-I recognizes short ssRNAs or blunt-ended dsRNAs (<300 bp) with 5'-triphosphate moiety while MDA5 detects long dsRNAs (>1000 bp) and RNA molecules with higher-order structures. For this reason, they have different roles in the host's antiviral response with RIG-I responding against paramyxoviruses, vesicular stomatitis virus (VSV), Japanese encephalitis virus, and influenza virus, whereas MDA5 functions against picornaviruses and noroviruses. There are some viruses that are recognized by both types of RLR, such as West Nile virus, dengue virus, and reovirus. In this case, RIG-I and MDA5 possibly recognize different sections of the same viral RNA. Meanwhile, LGP2 was observed to bind blunt-ended dsRNA of different lengths. RIG-I and MDA5 have two CARD domains, but this is absent in LGP2 so it lacks signal-transducing function. LGP2 is thought to regulate the activities of RIG-I and MDA5positively or negatively. It may competitively bind with RIG-I viral ligands or functions as a feedback inhibitor of RIG-I signaling by interfering with RIG-I oligomerization(Mogensen, 2009; Kato & Fujita, 2016; Rehwinkel & Gack, 2020; Kawai & Akira, 2009; Reikine *et al*., 2014; Prince *et al*., 2020). *In vitro* studies also show that LGP2 overexpression resulted to reduced IRF3 activation. In contrast, *in vivo* studies indicate that it enhances MDA5 signaling, possibly by facilitating the recognition of viral RNA by MDA5 (Reikine *et al*., 2014). It is also suggested that LGP2 regulates the survival and effector functions of the virus-specific CD8 T cell. Also, RIG-I and LGP2 have a repressor domain (RD) in their CTDs. The presence of RD explains why RIG-I overexpression results in the inability to initiate signaling in the absence of an activating ligand (Dixit & Kagan, 2013).

In the absence of their corresponding PAMPs, RLRs exist in the cytoplasm in a phosphorylated and inactive form. When bound by dsRNA, RIG-I dimerizes and shifts from close to open conformation, in which its CARD domains are released from being bound to the Hel domains enabling CARD-CARD interactions between RIG-I and the adaptor protein complex MAVS that is on the outer mitochondrial membrane. RIG-I is also dephosphorylated during the activation process, after which it is ubiquitinated by the E3 ubiquitin ligases Riplet and Tripartite motif 25 (TRIM25) (see Figure 2.24). The activated RIG-I then associates with other RIG-I/dsRNA complexes to form helical oligomers, which are linked together by the ubiquitin motifs. In contrast, unbound MDA5 assumes a more dynamic for maintaining equilibrium between open and close states, with the close form favored in the absence of the ligand. When MDA5 is activated, it assembles into ATP-sensitive filaments on the long dsRNA. The CARD domains on the outer region of the filaments interact with the CARD of MAVS (Reikine *et al*., 2014; Brisse & Ly, 2019).

When activated, MAVS polymerizes and forms fibrils, then it is polyubiquitinated and phosphorylated. The MAVS polymer serves as a platform that recruits different proteins, such as tumor necrosis factor receptor type-1 associated death domain (TRADD), RIP1, Fas-associated protein with death domain (FADD), tumor necrosis factor receptor-associated factors (TRAF2, TRAF3, TRAF3), caspase 8, and caspase 10 (see Figure

2.24). TRAF3 activates a complex made up of TANK-binding kinase 1 (TBK1), IκB kinase ε (IKKε), IKKγ, and TRAF family member associated NF-κB activator (TANK). This complex dimerizes and phosphorylates the transcription factorsIRF3 and IRF7 leading to their activation. Then, IRF3 and IRF7are translocated into the nucleus to induce the expression of IFNIs and other antiviral or immunoregulatory genes (Rehwinkel & Gack, 2020; Yong & Luo, 2018).TRAF2 and TRAF6 mediate the ubiquitination and subsequent activation of NEMO, which then stimulates the NF-κB pathway to induce the expression of pro-inflammatory cytokines (Yong & Luo, 2018).

A similar pathway activates MAPK, which then activates the transcription factor heterodimer ATF2/c-jun (Dixit & Kagan, 2013). Another mediator, MITA, located at the outer mitochondrial membrane, interacts with MAVS then recruits and activates IRF3 (Takeda & Akira, 2005). NF-κB and ATF2/c-jun are not sufficient to induce the expression of IFNβ. IRF3 and IRF7 are required for successful transcription (Dixit & Kagan, 2013).

Another adaptor molecule that stimulates IFN gene expression is STING that is located in the endoplasmic reticulum (ER). It interacts with RIG-I, but not MDA5, and MAVS and so it possibly enables interaction of the mitochondrion and the ER in terms of viral recognition and signaling. Such interaction is advantageous to the host's antiviral response since many viruses replicate in the membranous web connecting the ER and the mitochondrion (Mogensen, 2009).

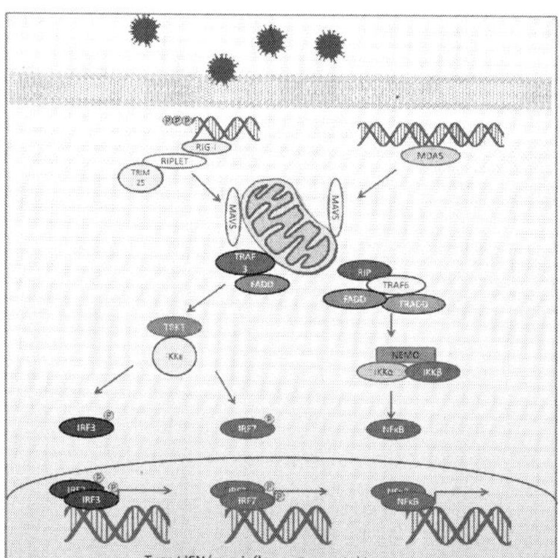

Figure 2.24. Retinoic acid-induced gene-I-like receptors (RLRs) like RIG-I and melanoma differentiation-associated protein 5 (MDA5) are activated upon

binding their viral RNA ligand. They interact with and activate mitochondria antiviral signaling protein (MAVS). MAVS recruits signaling proteins, such as tumor necrosis factor receptor-associated factor 3 (TRAF3), TRAF6, tumor necrosis factor receptor type 1-associated death domain (TRADD), receptor interacting serine/threonine protein kinase 1 (RIP1), and Fas-associated protein with death domain (FADD). TRAF3 activates TANK binding kinase 1 (TBK1) and IκB kinase ε (IKKε), which phosphorylate interferon regulatory factor 3(IRF3) and IRF7 causing them to dimerize. The resulting dimer goes into the nucleus to induce type I interferon (IFN) response. TRAF6 ubiquitinates NF-κB essential modulator (NEMO), which activates IκB kinase. This, in turn, activates NF-κB, which drives the expression of type I IFN and pro-inflammatory cytokines. (Source: Hui Yee Yong & Dahai Luo, Creative Commons License)

2.3.1.4. C-type lectin receptors

C-type lectin receptors(CLRs) act as lectins, which are proteins that bind carbohydrate groups. Similar to TLRs, some CLRs are transmembrane proteins located in the plasma membrane, while other members are soluble proteins that occur extracellularly. They contain at least one C-type lectin domain (CTLD), which is also called carbohydrate recognition domain (CRD) if it binds carbohydrates in a calcium-dependent manner, i.e., the Ca^{2+} ion acts as a bridge between the carbohydrate and the receptor (see Figure 2.25). The nature of this domain determines which type of carbohydrate is recognized. Thus, CLRscandetect sugar residues on foreign entities, such as glycans from the fungal and bacterial cell wall and viral envelope, but they can also recognize non-sugar molecules produced by damaged self-cells (Zaru, 2021). The bound ligands may be internalized, processed, and presented to T cells. They also induce signaling pathways that eventually activate NF-κB, IFNI, and inflammasomes. CLRs are expressed on the surface of DCs, macrophages, other myeloid cells, and epithelial cells. They are also expressed on APCs where they induce cytokine production and adaptive immune responses, such as T cell polarization, antibody production, and immunological memory formation (Mnich *et al*., 2020).CLRs are also involved in endocytosis, phagocytosis, cell adhesion, and tissue remodeling (Hutcheon *et al*., 2016).

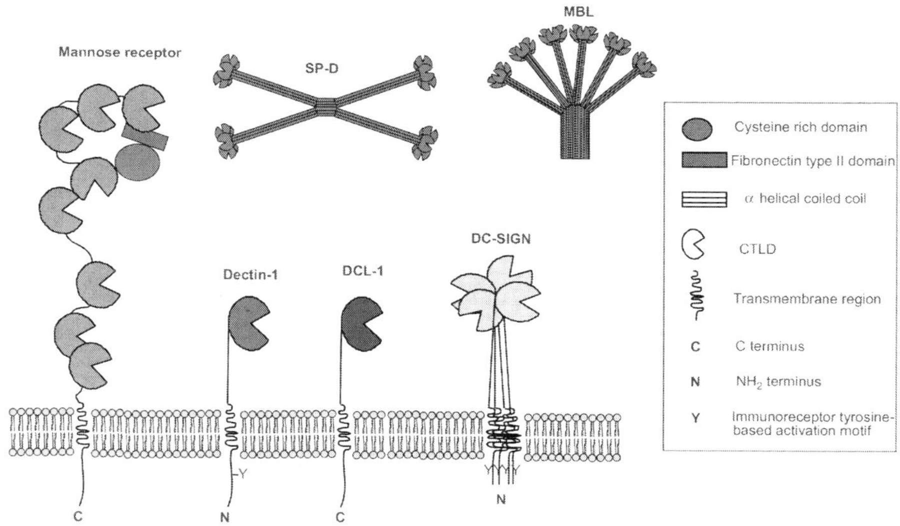

Figure 2.25. Transmembrane C-type lectin receptors (CLRs) contain at least one C-type lectin domain (CTLD), a stalk region, transmembrane domain, and an intracellular or cytoplasmic domain. Type I transmembrane CLRs have their N-terminus at the extracellular side, such as mannose receptor and DCL-1. Meanwhile, the N-terminus of type II is in the cytoplasm like in Dectin-1 and DC-SIGN. Thecytoplasmic domain of these two have a tyrosine (Y)-based signaling motif. The monomer of soluble CLRs like mannose-binding lectin (MBL) and surfactant protein-D (SPD) has CTLDs and α helical coils. Two MBL trimers can oligomerize to form a bouquet-like complex, while SP-D forms a cruciform dodecamer. (Source: Ann M. Kerrigan & Gordon D. Brown, Creative Commons License)

Depending on their topology, there are two types of CLRs expressed on APCs (see Figure 2.25). Type I transmembrane receptors have their N-terminal region pointing out of the cell and they have multiple CTLDs. Examples of this type include DEC 205 and macrophage mannose receptor (MMR). The type II transmembrane CLRs have their N-terminus oriented toward the cytoplasm and they have one CTLD in their C-terminus. These include dendritic cell-associated C-type lectin-1 (Dectin-1), Dectin-2, macrophage inducible C-type lectin (MINCLE), DC-specific ICAM-3 grabbing non-integrin (DC-SIGN or CD209), and CLEC9A. Both types have a stalk region, a transmembrane domain, and an intracellular domain with or without a signaling motif (Mnich *et al.*, 2020; Hutcheon *et al.*, 2016).

The type I and type II receptors can be further classified based on their conserved amino acid motifs in their CTLDs. These motifs influence their glycan specificity. One amino acid motif is EPN (glutamic acid-proline-asparagine), which recognizes glycans with equatorial 3- and 4-hydroxyl groups such as mannose, fucose, and N-acetylglucosamine (GlcNAc). This is observed in most human CLRs like DC-SIGN, langerin (CD207), and mannose receptor (MR or CD206). Another motif is QPD (glutamine-proline-aspartic acid) that preferentially binds glycans with axial 4-hydroxyl groups such as galactose and *N*-acetylgalactosamine (GalNAc). Macrophage galactose-type C-type lectin (MGL or CD301) is an example of a QPD CLR (Mnich *et al*., 2020).

Examples of soluble CLRsare mannose-binding lectin (MBL), surfactant proteins A (SP-A), and SP-D. MBL binds mannose, fucose, or GlcNAc residues on microbes then facilitates phagocytosis of the pathogen to enable antigen processing and presentation (Sue *et al*., 2018). Its structure is completely different from that of transmembrane CLRs (see Figure 2.25). Its basic functional unit is a homotrimer, with each monomer made up of an amino-terminal cysteine rich domain, a long collagenous domain, α-helical coiled coil, and a CTLD. An oligomer composed of two of these homotrimers forms a bouquet-like structure. Oligomerization enables MBL to bind repetitive carbohydrate moieties that are found on microbial surfaces. SP-A and SP-D have a basic structure similar to MBL. SP-D forms dodecamers with a cruciform shape. They are synthesized in the lungs, where they are the main protein constituent of the pulmonary surfactant. They were observed to recognize and bind a variety of carbohydrates on pathogens and then interact with surfactant receptors on phagocytic cells (Kerrigan & Brown, 2009).

The ability of CLRs to interact with signaling molecules or the presence of certain signaling motifs in their cytoplasmic tails determine whether they will have an activating or inhibiting effect (see Figure 2.26). There are at least four types of intracellular signaling motifs. One is immunoreceptor tyrosine-based activation motifs (ITAMs), which have YXXL (tyrosine-amino acid-amino acid-leucine) tandem repeats that may interact with ITAM-containing adaptor proteins like the Fc receptor γ (FcRγ) chain. Another type is hemi ITAM (HITAM), which has a single tyrosine within a YXXL motif. The tyrosine residues of ITAM and hemiITAM are phosphorylated when an activating CLR binds its ligand. The phosphorylated ITAM facilitates the recruitment of spleen tyrosine kinase (SYK) and PKσ that induce the assembly and activation of the CARD9/ B cell lymphoma 10

(Bcl10)/mucosa-associated lymphoid tissue lymphoma translocated gene 1 (Malt1) complex. This results in NF-κB activation and ROS production that leads to activation of inflammasomes. Dectin-1 is an interesting CLR with ITAM since it has an alternative pathway for regulating NF-κB activation that is mediated not by SYK but by RAF-1 kinase (Hutcheon *et al*., 2016; Chiffoleau, 2018; Drouin *et al*., 2020).

In contrast, the tyrosine residue of immunoreceptor tyrosine-based inhibitory motifs (ITIMs) is phosphorylated when a certain ligand binds an inhibitory CLR. This is followed by the recruitment and activation of protein tyrosine phosphatases, such as the Src homology region 2 domain-containing phosphatase 1 (SHP1) or SHP2 resulting in dephosphorylation of substrates. This leads to negative regulation of cellular activation pathways mediated by other immune receptors like TLR8-mediated IL12 and TNF production or TLR9-induced IFNα and TNF synthesis. This serves as a checkpoint to prevent uncontrolled immune responses that may have harmful consequences. The last type is tyrosine-independent motif, which is found in some CLRs with no ITAM or ITIM domains so that their signaling pathway is either uncharacterized or an alternative one. Nevertheless, a CLR may have both activating or inhibiting action depending on the nature of the ligand or the environment (Hutcheon *et al*., 2016; Geijenbeek & Gringhuis, 2009; Chiffoleau, 2018; Drouin *et al*., 2020).

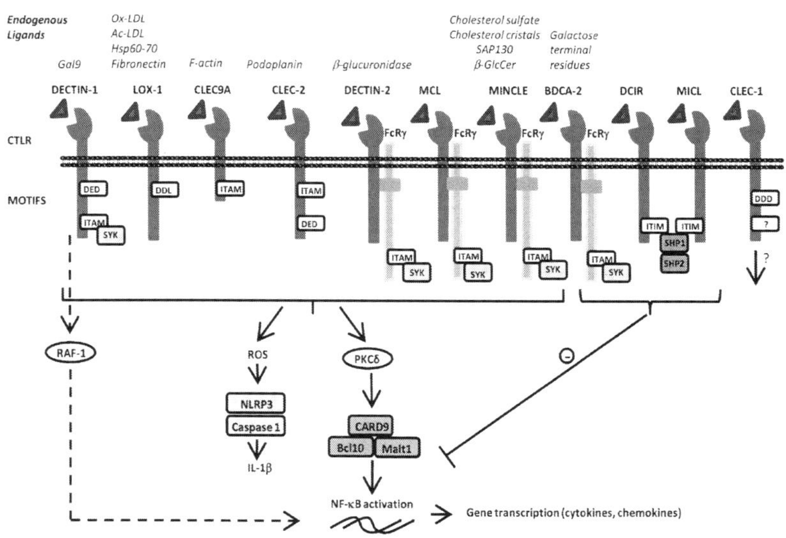

Figure 2.26. C-type lectin receptors (CLRs) can have either an activating or inhibiting effect depending on the signaling motif they have in their cytoplasmic

domain. When the immunoreceptor tyrosine-based activation motif (ITAM) is activated, the SYK kinase family is recruited and activated. Then PKσ activates the CARD9-Bcl10-Malt1 complex that triggers the NF-κB pathway and subsequent expression of cytokines and chemokines. The ITAM of Dectin-1 may induce an alternative pathway that is mediated by RAF-1 instead of SYK.SYK also induces reactive oxygen species (ROS) generation and inflammasome activation resulting in interleukin-1β (IL-1β) production. Activation of immunoreceptor tyrosine-based inhibition motif (ITIM) triggers the recruitment and activation of protein tyrosine phosphatases like SHP1 or SHP2. These dephosphorylate the motifs resulting in the inhibition of signaling pathways that are mediated by other immune receptors. Other CLRs like CLEC1 have unknown signaling motifs. (Source: Elise Chiffoleau, Creative Commons License)

Apart from recognizing PAMPs on pathogens, CLRs detect DAMPs from damaged or dead cells (see Figure 2.27). This function complements the detection of pathogens. CLEC9A on dendritic cells binds to exposed F-actin on dying cells, which are eventually phagocytosed. Its HITAM, when phosphorylated, interacts with SYK. The engulfed cell is then directed to a non-degradative recycling endosome to allow cross-presentation of the dead cell-associated antigens to CD8T cells. CLEC9A HITAM also activates SHP1 to exert inhibitory feedback mechanisms in order to control neutrophil recruitment and prevent further tissue damage. MINCLE expressed by macrophages recognizes ligands released by necrotic cells, such as nuclear sin3A associated protein 130 (SAP130), β-glucosylceramide, cholesterol sulfate and crystals. When MINCLE is bound, its ITAM is activated. As a result, the downstream NF-κB, MAPK or AP1signaling pathway is activated resulting in enhanced pro-inflammatory responses. Alternatively, MINCLE can also signal through PLCγ2 to induce the calcineurin/nuclear factor of activated T cells (NFAT) pathway, which also triggers the transcription of chemokine and cytokine genes. In the presence of a tumor, MINCLEtriggers macrophage-induced immune suppression and cancer progression. When CLEC8A or LOX-1 is triggered by oxidized low density lipoprotein (LDL), it results in ROS generation that activates inflammasomes and induces apoptosis signals. CLEC8A also promotes tumor immune escape and metastasis. Most myeloid cells express CLEC12A that binds monosodium urate crystals produced during cell death. This triggers its ITIM to recruit SHP1 and SHP2 to counteract activating signals. This mode of action serves to prevent harmful inflammation (Drouin *et al*., 2020).

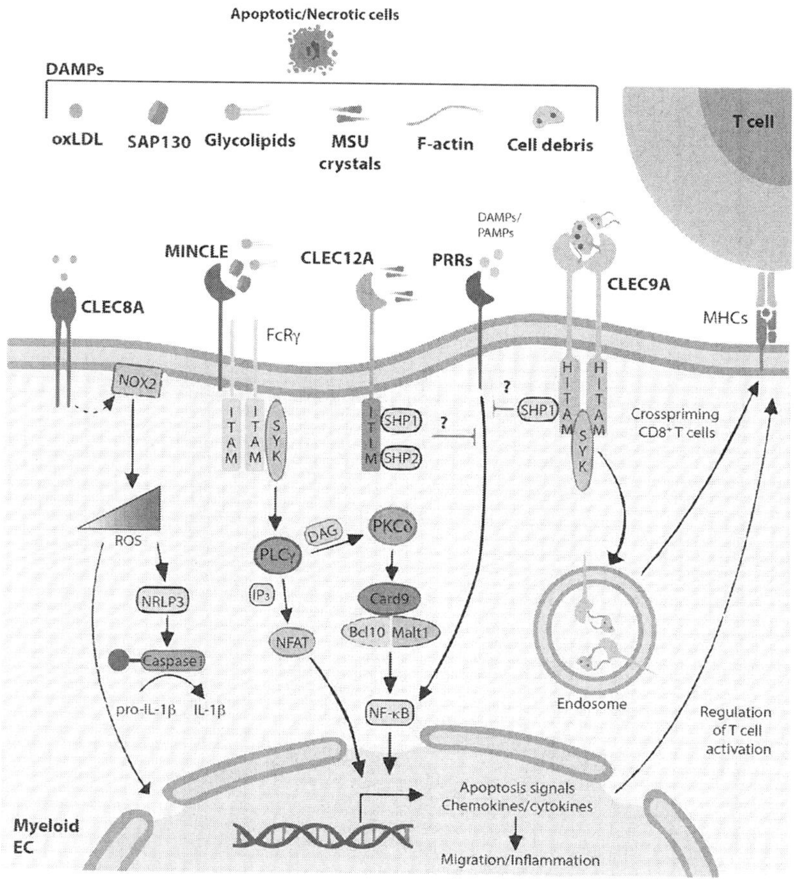

Figure 2.27. C-type lectin receptors (CLRs) recognize damage-associated molecular patterns (DAMPs). CLEC8A binds oxidized low density lipoproteins (oxLDL) leading to reactive oxygen species (ROS) generation. This triggers IL-1β production. Sin3A associated protein 130 (SAP130) from necrotic cells is recognized by MINCLE resulting in the activation of immunoreceptor tyrosine-based activation motif (ITAM). This leads to the recruitment and activation of the SYK kinase family and the subsequent activation of the CARD9-Bcl10-Malt1 complex, which induces NF-κB signaling activation, or PLCγ2 that activates the NFAT signaling pathway. Both pathways induce the transcription of cytokine and chemokine genes. CLEC12A interacts with monosodium urate (MSU) crystals, which activates the immunoreceptor tyrosine-based inhibition motif (ITIM). This triggers the recruitment and activation of protein tyrosine phosphatases, SHP1 and SHP2, that dephosphorylate activation motifs. This inhibits cellular activation mediated by other pattern recognition receptors

(PRRs). CLEC9A detects F-actin on cell debris. Its hemi-ITAM (HITAM) is involved in facilitating $CD8^+$ T cell cross-priming. CLEC9A can also activate SHP1 to restrain excessive immune response. (Source: Marion Drouin *et al*., Creative Commons License)

2.4 TYPES OF INNATE IMMUNE RESPONSES

There are four types of innate immune systems (see Figure 2.28). These are the anatomic barriers (skin and mucous membrane), physiologic barriers (temperature, low pH, and chemical mediators), endocytic and phagocytic response, and inflammatory response (Marshall *et al*., 2018). As previously mentioned, the innate immune responses are not mutually exclusive from the adaptive immune system, but the two types of immunity complement each other.

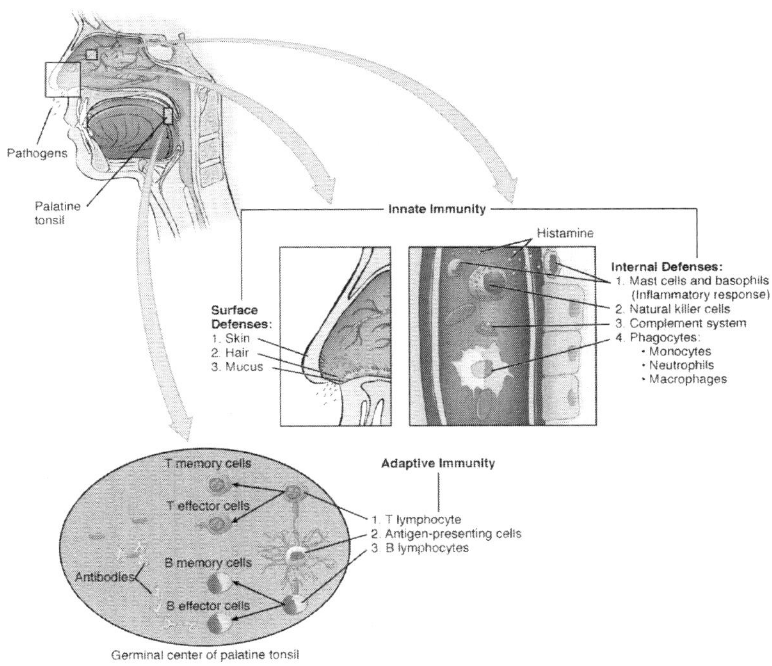

Figure 2.28. The innate immune system comprises of the anatomic or surface defenses, physiologic defenses as well as internal defenses made up of phagocytic cells and cells involved in the inflammatory response. It complements adaptive immunity in order to enhance the over-all immune response. (Source: Lindsay M. Biga *et al*., Creative Commons License)

2.4.1 Anatomical and Surface Defenses

Epithelial surfaces such as the skin and the respiratory and digestive lining serve as a physical barrier that prevents molecules from the external environment from getting inside the body and maintains inner homeostasis. Adjacent cells of the epithelium are connected by tight junctions, which prevent potential pathogens and their products from inserting themselves between cells. The slimy mucus layer in the interior epithelial surfaces prevents microbes from adhering to the epithelium. The beating action of the cilia on the epithelial cells also facilitates the removal of pathogens and other foreign molecules (Alberts *et al*., 2002).

The epithelial surfaces are more than physical obstructions. They also function as chemical and biological (or physiological) defenses. Apart from the barrier properties of the epithelium, injured or infected epithelial cells can produce chemotactic signals that initiate the inflammatory response. Moreover, certain epithelial cells also produce antimicrobial substances, such as the lysozyme in tears, saliva, and respiratory secretions, lactoferrin, defensins, and peroxidase. It was also demonstrated that epithelial cells express on their cell membranes an antimicrobial substance called bactericidal permeability-increasing (BPI) protein. This protein can bind to the lipopolysaccharide layer on the outer membrane of Gram-negative bacteria and proceed to kill them (Ganz, 2002).

The response to bacterial lipopolysaccharide is an example of PAMP recognition. It is recognized by the Toll-like receptor (TLR)known as TLR4, which is located on the basolateral surface of epithelial cells or on macrophages and other immune cells found below the epithelial surface. Other types of TLRs found in the same location include TLR2 that recognizes peptidoglycan and mycobacterial glycolipids and TLR5 that recognizes bacterial flagellin. These receptors are activated when they bind the microorganisms that penetrate the epithelial barrier in an injured area (Ganz, 2002).

The epithelium does not only defend the body from harmful microbes, but it also hosts many beneficial microorganisms. These are a population of approximately 10^4-10^{12}microorganismsmade up of at least 2,000 archaeal, bacterial, fungal, and viral species that reside in the small intestine, colon, skin, nasal and oral cavities, and the female reproductive tract (Moens & Veldhoen, 2012; Abdallah *et al*., 2017; McGhee & Fujihashi, 2012). This microbial community contributes to health and wellbeing by facilitating digestive and anabolic functions and by preventing pathogen colonization.

These "good" microbes may compete with the pathogens for nutrients and attachment sites to epithelial cells. Some can produce bacteriocins or induce antimicrobial substances to kill their harmful competitors (Moens & Veldhoen, 2012).

2.4.1.1 Immunity in the skin epithelial barrier

The skin is the body's largest organ, and it is critical in providing protection from environmental pressures such as mechanical injury, extreme temperatures, ultraviolet (UV) rays, allergens, toxins, allergens, and of course pathogenic microorganisms. It is made up of a network of various cell types including endothelial cells, fibroblasts, neurons, adipocytes, and epithelial cells. The skin also has resident and recruited immune cells that prevent microbial invasion and repair the barrier (Kobayashi *et al.*, 2019). The epidermis has resident immune cells and structural cells with immune functions, while other specialized immune cells like DCs, macrophages, and T cells are located in the inner dermis. These immune cells closely interact with the neighboring cells like keratinocytes and fibroblasts. Skin immune response also depends on the dermal vessels and lymph nodes that drain the skin (Matejuk, 2018).

2.4.1.1.1 Skin cells involved in immunity

The major immune cells in the skin express various types of PRRs, both the transmembrane and cytosolic types. Most of the immune cells are in the dermis and these interact with epithelial cells resulting in an effective immune response (Abdallah *et al.*, 2017).

Keratinocytes make up several layers of the epidermis, constituting 95% of epidermal cells and playing an important role in the skin's structural integrity (see Figure 2.29). Technically, they are not immune cells, but they serve as immune sentinels and initiators of immune response and are involved in both innate and adaptive immune systems. The outermost corneal layer (stratum corneum) in the epidermis is made up of dead keratinocytes that are linked together by desmosomes to form a hydrophobic matrix. This layer serves as a mechanical barrier against irritants, allergens, and physical force. On the other hand, the keratinocytes in the granular (stratum granulosum), spinous (stratum spinosum), and basal (stratum basale) layers can detect external stimuli like UV rays and chemicals, which activate inflammasome-mediated inflammatory signaling leading to the production of IL1β (Matejuk, 2018; Afshar & Gallo, 2013; Abdallah *et al.*, 2017).

They also harbor PRRs like TLRs, which recognize microbial ligands and when activated, induce the production of cytokines, chemokines, growth factors, and AMPs as well as promote Th1 responses and IFN production. Extracellular bacterial or fungal pathogens stimulate the production of IL17, CCL20. CXCL9 or CXCL10 to recruit and activate Th17 or Th1 cells. In an infected state, IL17 can activate AMP synthesis by keratinocytes. On the other hand, intracellular bacteria, protozoa, and viruses trigger IL12, IL15, and IFNγ-mediated responses. Upon stimulation with IFNγ, keratinocytes can express MHCII so that they can function as APCs for T cells. The interaction between T cells and keratinocytes is significant in responding to local antigens and in maintaining self-tolerance. Gram-negative bacteria can trigger increased synthesis of TNF, IL1, and other cytokines that stimulate the increased production of CXCL1, CXCL2, CXCL3, CXCL5, and CXCL8 that recruit neutrophils into infected sites. On the other hand, helminth infection activates IL25, IL33, CCL17, and CCL22 production by keratinocytes to drive Th2-mediated responses (Matejuk, 2018; Abdallah *et al.*, 2017; Guttman-Yassky *et al.*, 2019).

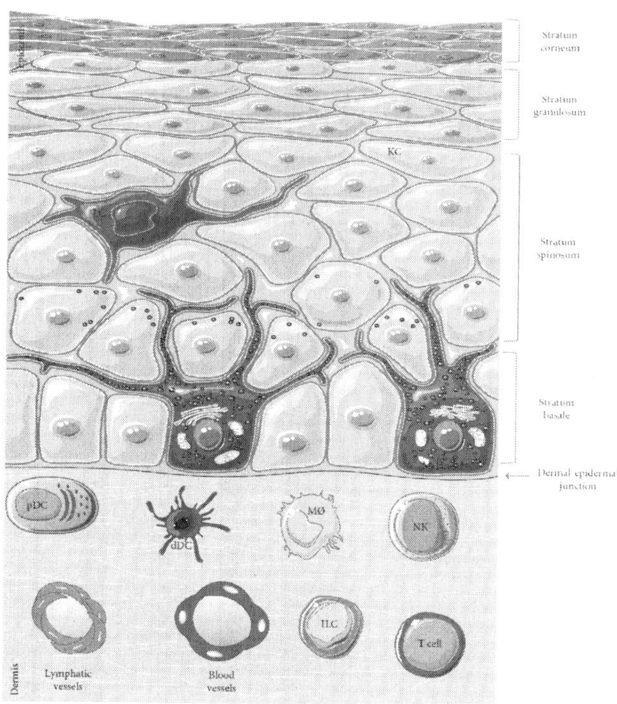

Figure 2.29. The skin's capacity to protect the body from stressors in the external environment is due to its complex structure and its inherent immune system.

The outermost epidermis is made up mostly of keratinocytes (KCs), which are arranged in several sublayers or strata, namely the most external stratum corneum, followed by stratum granulosum, stratum spinosum, and stratum basale. Langerhans cells (LCs) are found in the inner strata of the epidermis, while the melanin-producing melanocytes (blue-colored cells) occur in the innermost stratum basale. The dermis is home to blood and lymphatic vessels as well as a variety of immune cells, such as plasmocytoid dendritic cells (pDCs), dermal dendritic cells (dDCs), macrophages (MØs), natural killer cells (NKs), innate lymphoid cells (ILCs), and T cells. (Florence Abdallah *et al.*, Creative Commons License)

Melanocytes are epithelial cells located in the basal layer of the epidermis, near the dermal-epidermal junction (see Figure 2.29). They produce the skin pigment melanin that provides color to the skin and protects keratinocytes from the harmful effects of UV radiation. Upon coming in contact with pathogens, they produce cytokines, such as IL1β, IL6, and TNFα, and chemokines that recruit leukocytes like CCL2, CCL3, and CCL5.PAMPs can also activate the expression of TLRs (e.g., TLR1, TLR2, TLR3, TLR4, TLR6, TLR7, TLR9) on melanocytes driving them to produce IL6 and IL8. Melanocytes may also express IFNI so that they are a part of the first line of defense against viral infections in the skin. They are also targets of skin pathogens like alphavirus, varicella-zoster virus, parvovirus, *Mycobacterium leprae*, and *Leptospira*. Thus, change in skin pigmentation is an important symptom observed in the diseases caused by these infectious agents (Abdallah *et al.*, 2017; Quaresma, 2019).

Another cell type that interacts with keratinocytes for immunity purposes is the fibroblasts, which are the main cell component of connective tissues. They secrete collagen and fibronectin, which are components of the extracellular matrix, as well as proteinases that are involved in tissue remodeling. They are not considered immune sentinels although they secrete various cytokines, chemokines, and growth factors. The secreted cytokines may facilitate the fibroblast-keratinocyte interaction to generate skin immunity and enhance the wound healing process (Abdallah *et al.*, 2017).

An immune cell in the epidermis is the Langerhans cell, a type of DC located in the epidermis (see Figure 2.29). This makes the Langerhans cells the first immune cells to come in contact with pathogens invading the skin. They make up 2-4% of the cells in the epidermis and they have a 53 to 78-day half-life. When the Langerhans cells in the spinous layer are stimulated, their dendrites elongate, extending into the epidermal tight junctions to reach the stratum corneum and capture antigens without disturbing the epithelial barrier. They will then migrate into draining lymph nodes in the dermis where naïve T cells are activated. Langerhans cells express PRRs, on their plasma membrane, namely the C-type lectin receptors CD1a, Fcγ receptors, and Fcε receptors. These recognize mannosylated PAMPs on a broad range of pathogens, which leads to receptor-mediated endocytosis of the pathogen and its transport to the Birbeck granule where it may participate in antigen processing (Abdallah *et al.*, 2017). Langerhans cells produce IL10, induce CD4 regulatory T (Treg) cells, and suppress CD8 T cells. Thus, they function to induce inflammatory responses and maintain immunologically tolerant or non-inflammatory states (Matejuk, 2018).

Macrophages in the dermis are critical in the immediate elimination of pathogens via phagocytosis and in the control of inflammation (see Figure 2.29). They accomplish these in three phases. In the first phase, the FcγRI of macrophages recognizes the crystallized fragment (Fc) of IgG-covered microbes inducing antibody-dependent cell cytotoxicity (ADCC) and phagocytosis. These result in the destruction of pathogens. Another receptor on the macrophages, C3bR, detects microbes coated with the complement C3b and initiates lysis or phagocytosis. The second phase depends on the secretion of pro-inflammatory mediators like certain cytokines (e.g., TNFα, IL6, and IL1β), ROS, and nitric oxide. The final phase suppresses inflammation via anti-inflammatory mechanisms triggered by TGFβ and lipid mediators (Abdallah *et al.*, 2017).

Dendritic cells (DCs) are the predominant immunocompetent cells in the dermis (see Figure 2.29) and are also found close to the hair follicles. They coordinate innate and adaptive immunity in the skin. Adaptive immunity is necessary when innate immune responses failed to sufficiently eliminate the pathogens. The main role of dermal dendritic cells (dDCs) is to detect and provide protection against pathogens by mediating inflammatory responses. There are two types of dDCsbased on the expression of blood dendritic cell antigen (BDCA), namely BADCA1 and BDCA3. BDCA1 dDCs are immature cells that express TLR2 and TLR4, produce low amounts of IL10, and do not induce Treg cells. They have relatively weak Tcell stimulatory potential, unless activated by pro-inflammatory cytokines. BDCA3 dDCs, on the other hand, produce epithelial cell adhesion molecules and high amounts of IL10. They promote differentiation of CD25 Treg cells and have the ability to stimulate B cells to become IgM-secreting plasma cells (Matejuk, 2018; Guttman-Yassky *et al.*, 2019). The dDCs are also migratory since they can capture skin antigens by phagocytosis, mature to express

MHC II and act as APC, and then move to draining local lymph nodes. Then, they interact with lymphocytes and drive the synthesis of cytokines and costimulatory molecules. The activation of co-stimulatory molecules facilitates the activation of T cells (Abdallah *et al.*, 2017; Quaresma, 2019). Human skin DCsalso express the antigen-presenting molecules CD1a, CD1b, and CD1c that allow lipid presentation to T cells (Kobayashi *et al.*, 2019). It is quite rare to find plasmacytoid DCs (pDCs) in the human skin. They have TLR7 and TLR9 that play a role in responding to viral infection by triggering the release of large amounts of IFNα (Abdallah *et al.*, 2017; Quaresma, 2019).

There are mast cells located in the upper dermis. Their histamine content makes them known as allergy cells, but they are also involved in wound healing, skin inflammation, angiogenesis, immune tolerance, and cancer. They have TLRs that facilitate inflammatory responses and they also function as APCs via their MHCI and MHCII. The human skin has mostly the TC type (i.e., tryptase positive, chymase positive) of mast cells. These have the highest proteinase content, namely tryptase, chymase, carboxypeptidase, and a cathepsin G-like proteinase. Tryptase cleaves fibronectin and extracellular matrix proteins so that immune cells like neutrophils, mononuclear cells, and lymphocytes can infiltrate the epidermis and initiate inflammatory pathways. Similarly, chymase also attracts and activates various immune cells. It also cleaves precursor interleukins in order to produce active forms of IL1β and IL18. On the other hand, both tryptase and chymase degrade several pro-inflammatory cytokines and chemokines. As a result, they also downregulate immune responses. Mast cells can also enhance the activation of other immune cells like Treg, trigger Th2 cell-mediated responses, activate IgE production by B cells, and promote the migration and maturation of professional APCs. They have been shown to produce the anti-inflammatory cytokine IL10 so that they inhibit skin inflammatory reaction in cases of contact hypersensitivity and UV-irradiation (Matejuk, 2018; Guttman-Yassky *et al.*, 2019).

The skin contains about 20 billion T cells, almost twice the number present in the blood. There are both resident and transient T cells in the skin. There are memory T cells that are responsible for providing the first line of defense against local infection. Meanwhile, Treg cells, which are a subtype of T cells, constitute 5-10% of all resident T cells in the skin. They circulate between the skin and lymph nodes in the normal state and during an immune response. Treg cells in normal skin produce a high quantity of anti-inflammatory cytokines, such as IL10, IL35, and TGFβ (Matejuk,

2018; Guttman-Yassky *et al.*, 2019). Other resident T cells are the γσ T cells and NKT. The γσ T cells produce IL17, growth factors that promote wound healing, and AMPs. NKT cells recognize bacterial and self-derived glycolipids. Mouse models show that, when activated, these cells maintain high levels of TNFα and increase DC migration from the skin to draining lymph nodes (Matejuk, 2018).

2.4.1.1.2 Mechanism of skin immunity

When the skin epidermis is wounded, microbes may be introduced to the site of injury (see Figure 2.30). The PAMPs on these foreign invaders as well as DAMPs from damaged cells are recognized by various types of PRRs on the cell surface and endosomes of keratinocytes. For instance, TLR2, TLR6, and NOD2 detect *Staphylococcus aureus*, while TLR2, TLR3, TLR7, TLR8, and TLR9 recognize many viruses. Fungi such as *Candida albicans* activate CLRs, TLR2, TLR4, and TLR9. Activation of PRRs triggers the inflammatory cascade so that pro-inflammatory cytokines, such as IL1β, , IL18, TNFα, and IFNγ are produced by keratinocytes. In response to these cytokines, adhesion molecules are expressed on dermal endothelial cells and MHC II are expressed on keratinocytes and Langerhans cells. In addition, IL1 can stimulate the expression of ICAM-1. CXC chemokines like IL8 interact with the adhesion molecules and recruit neutrophils, which initiate phagocytosis of the pathogens and removal of cell debris in the wounded area. Pro-inflammatory M1 macrophages continue performing phagocytosis and amplify the inflammatory response. Then anti-inflammatory cytokines like TGFβ convert M1 macrophages to the anti-inflammatory M2 phenotype, which promotes wound repair and closure (Coates *et al.*, 2018; Matejuk, 2018).

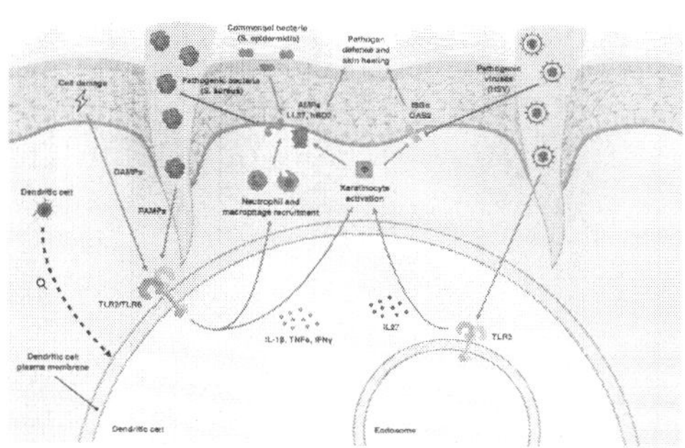

Figure 2.30. Pathogenic microorganisms can penetrate the epidermis when the skin is injured. The pathogen-associated molecular patterns (PAMPs) in these

microbes (e.g., *Staphylococcus aureus*) and the damage-associated molecular patterns (DAMPs) from damaged cells are recognized by Toll-like receptors (TLRs), such as TLR2/TLR6. This induces the dendritic cells to express pro-inflammatory cytokines like IL-1β, TNFα, and IFNγ. These recruit neutrophils and macrophages to the site of injury and promote AMP production, such as LL37 and hBD2. TLR3 is specifically activated by viral PAMPs, resulting in IL27 production. IL27, in turn, induces the expression of antiviral interferon-stimulated genes (ISGs), such as oligoadenylate synthetase 2 (OAS2), by keratinocytes. Pathogenic bacteria inhibit AMP production, while commensal bacteria like *Staphylococcus epidermidis*, promote it. Viruses may also inhibit ISG production. (Source: Margaret Coates *et al.*, Creative Commons License)

On the other hand, certain PRRs like TLR3 specifically detects viral PAMPs resulting in the IL27-mediated expression of antiviral ISGs. An example is oligoadenylate synthetase (OAS), which binds viral dsRNA and then activates an intracellular latent RNase (RNaseL) so that viral RNA is digested (Coates *et al.*, 2018).

AMPs in the skin are produced by keratinocytes mainly as well as by infiltrating immune cells like neutrophils and macrophages and by commensal microorganisms, but its production may be inhibited by pathogenic bacteria (see Figure 2.30). AMPs are primarily produced in the stratum granulosum, packaged into lamellar bodies, and then transported to the stratum corneum. These are also present in skin secretions like saliva and sweat. Some AMPs are constitutively expressed in epithelial cells, while others are normally present at low levels then significantly increase during inflammation. Their production is triggered by the activation of receptors that recognize lipopolysaccharides from Gram-negative bacteria, lipoteichoic acid and peptidoglycan from Gram-positive bacteria, mannans from yeast and fungi, and nucleic acids from pathogens and self-cells. These receptors include TLRs, mannose receptors, and helicases. Some pro-inflammatory cytokines like IL17 and IL22 can upregulate AMP expression. Correspondingly, it was observed that the cholinergic anti-inflammatory pathway decreases AMP production (Coates *et al.*, 2018; Matejuk, 2018; Afshar & Gallo, 2013).

Most AMPs have a net negative positive charge that enables them to electrostatically interact with the negatively charged phospholipids in the cell membrane of bacteria and anionic components of fungi and viruses. The peptides associate with the lipid head groups of the phospholipid bilayer leading to the formation of transmembrane pores. These pores disrupt and destabilize the bacterial cell membrane resulting in lysis of the bacterium. AMPs preferentially target replicating bacteria, especially those at the site of cell division. AMPs may also have an immunomodulatory function by influencing cell signaling pathways (Afshar & Gallo, 2013).

Cathelicidins and defensins are the well-characterized AMPs. Cathelicidin resides predominantly in granules of the outer epidermis and in the extracellular spaces in the stratum corneum. They can neutralize Gram-negative bacteria, Gram-positive bacteria, fungi, and viruses as well as promote the repair of skin wounds by transactivating epidermal growth factor receptor, which induces the migration of keratinocytes. The human cathelicidin **LL37** is induced by injury, bacterial PAMPs, pro-inflammatory cytokines, retinoic acid, and vitamin D_3. It protects from Group A *Streptococcus* and neutralizes lipopolysaccharide endotoxin. It also triggers chemotaxis of neutrophils, monocytes, mast cells, and T cells to the site of injury and infection, then stimulates angiogenesis and wound healing (see Figure 2.30). LL37 also induces histamine release from mast cells and intracellular Ca^{2+}mobilization as well as enhances the phagocytic capacity of DCs by altering the expression of phagocytic receptors. It associates with extracellular self-DNA fragments allowing them to enter plasmacytoid DCs, which leads to TLR9 activation and ultimately to IFNI synthesis. On the other hand, it inhibits macrophages from releasing pro-inflammatory cytokines as well asTLR4-mediated maturation of DCs and cytokine release (Coates *et al.*, 2018; Matejuk, 2018; Afshar & Gallo, 2013).

β-defensins are the major defensin family produced in the skin of which there are three types in humans, namely human β-defensin 1 (hBD1), hBD2, hBD3, and hBD4. These are produced by keratinocytes, sebocytes, and sweat glands. Keratinocytes constitutively express hBD1 and hBD2, while hBD3 and hBD4 levels are low in the normal state and upregulated by infection, injury, and pro-inflammatory cytokines. In addition, *Staphylococcus epidermidis*, a member of the skin microflora, upregulates hBD2 and hBD3 expression via TLR2-induced p38 mitogen-activated protein kinase signaling. In terms of antimicrobial activity, hBD1 and hBD2 can protect strongly against Gram-negative bacteria, such as *Escherichia coli* and *Pseudomonas aeruginosa*, but they are not active against Gram-positive bacteria. Still, hBD2 prevents secondary bacterial infection in cases of atopic dermatitis. Meanwhile, hBD3 exhibits broad-spectrum activity against Gram-negative and Gram-positive bacteria as well as fungi. Human β-defensins also influence other components of the skin immune system. For instance, hBD2attracts DCs and memory T cells and promotes the production of histamine by mast cells. In addition, hBD2, hBD3, and hBD4 stimulate keratinocytes to upregulate their expression of pro-inflammatory cytokines and chemokines. Keratinocytes are also triggered by β-defensins to migrate and proliferate, thus promoting wound repair (Coates *et al.*, 2018; Matejuk, 2018; Afshar & Gallo, 2013).

Dermcidin is an AMP constitutively produced as a small precursor protein by sweat glands and it accounts for the antimicrobial property of sweat. Once the precursor is secreted into the sweat, it is catalytically cleaved in order to be activated. Dermcidin-derived AMPs do not induce membrane pores, but they bind to the bacterial cell envelope, resulting in decreased gene expression. They demonstrate antimicrobial activity against *S. aureus*, *E. coli*, and *C. albicans* (Coates *et al.*, 2018; Matejuk, 2018; Afshar & Gallo, 2013).

Psoriasin is the most hydrophobic AMP in the skin, which is secreted with sebum lipids. Great quantities of psoriasin accumulate in sebaceous glands and sebaceous skin epidermis.*E.coli* triggers its expression in keratinocytes (Afshar & Gallo, 2013).

The skin hosts beneficial microorganisms that form the skin microbiota. This works with the skin immune system to provide protection against pathogens via AMP production, inhibition of pathogen adhesion, toxin degradation, and activation of the skin immune pathways as well as to maintain tolerance to resident microorganisms. The skin microbiota occurs in all skin layers but 25% of microorganisms grow in the dermis at the level of sebum glands and through hair follicles. Some of these are resident microorganisms that have been transmitted from the mother at birth or acquired from daily exposure to the environmental factors (e.g., animals, plants, other people, chemicals, climates). Others are transient microorganisms that are acquired due to changes in one's surroundings, like when traveling, and then removed once the conditions are back to the usual state. Each person, therefore, has a unique and specific skin microbiota that is developed during infancy and then stabilized during adulthood (Abdallah *et al.*, 2017).

S. epidermidis is a resident beneficial microorganism that is most frequently isolated from human epithelia. It produces a group of phenol-soluble modulins that exhibit antimicrobial action against pathogenic bacteria, such as group A *Streptococcus*, and restrain the pathogenicity of *Staphylococcus aureus*(see Figure 2.31). In mice, it triggers an inflammatory response mediated by IL1, IL17, and IFNγ in order to protect against parasites like *Leishmania major*. It also releases a certain substance that enhances keratinocytes to produce AMP and activates TLR2-mediated signaling. The lipoteichoic acid in the cell wall of *S. epidermidis* and other commensal gram-positive bacteria can also increase mast cell activity against vaccinia viruses and inhibit inflammation induced by *Propionibacterium acnes* (Coates *et al*., 2018; Abdallah *et al*., 2017). On the other hand, *Malassezia* spp. is one of the most dominant fungi in the skin that protects against bacteria and other fungi. It metabolizes external lipids to produce short fatty acids like azelaic acid, which exhibits antimicrobial activity at normal skin pH (Abdallah *et al*., 2017).

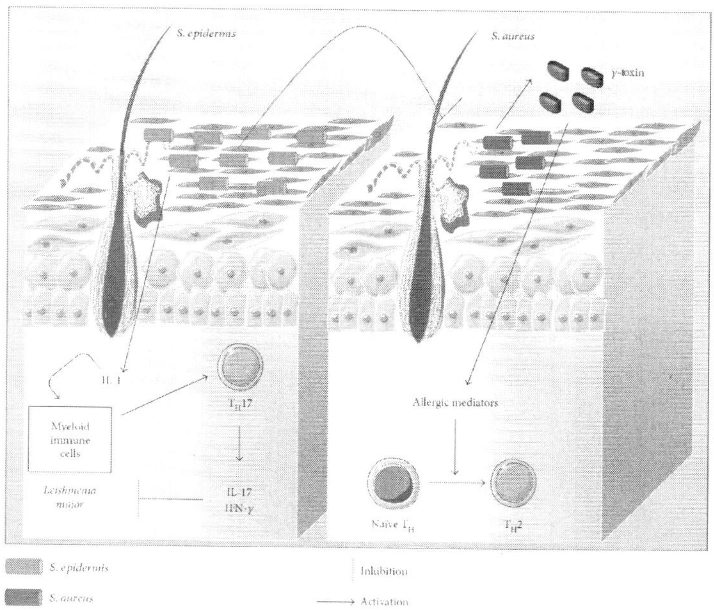

Figure 2.31. The skin immune system interacts with skin microbiota to protect against pathogens and to tolerate commensal microorganisms. In a healthy state, *Staphylococcus epidermidis* can control *Staphylococcus* pathogenicity. It stimulates T cells to produce IL17 and IFNγ, therefore inhibiting *Leishmania major* proliferation. *S. aureus* produces σ toxins that trigger local allergic responses in the skin activating Th2-mediated inflammatory response. (Source: Florence Abdallah *et al*., 2017)

When the balance within the skin microbiota is disrupted, its alliance with the skin immunity can be terminated so that certain skin diseases may develop, such as atopic dermatitis, acne, and rosacea. *S. aureus* is a transient member of the skin microbiota and 30-50% of the human population are healthy carriers of the bacterium. However, it becomes a pathogen as soon as it is left uncontrolled by other members of the resident microbiota It produces σ toxins that cause local allergic responses in the skin, which may prevent wound healing and cause epithelial barrier deterioration (see Figure 2.31). When the epidermal chemistry (i.e., pH, pathological sweat secretion) is altered, *Malassezia* spp. gains in pathogenicity and releases lipases, phospholipases, and indoles. These disrupt the epithelial barrier so that immune deregulation and diseases ensue (Abdallah *et al*., 2017).

Tissue-resident memory T cells (Trms) are formed after the clearance of infection and since they are long-lived cells, they confer long-term skin immunity. They accumulate at the site of inflammation as well as at other regions of the skin that were not involved in the previous infection. The keratinocytes are the main suppliers of the cytokines IL7 and IL15, which are essential for Trm generation and maintenance (Kobayashi *et al*., 2019).

2.4.1.2 Immunity in the gastrointestinal tract epithelium

The gastrointestinal tract is under constant threat from invading pathogens, chemical toxins, and food allergens. The acidity of the stomach, the mucus lining, the antimicrobial enzymes, and the innate immune system in the gut can control microbial invasion. Similar to the skin, the gut also has many beneficial microorganisms. They colonize the gastrointestinal tract immediately after birth. They are made up of various species of microorganisms, including viruses, but there are four predominant bacterial phyla in the gut microbiota, namely *Firmicutes*, *Bacteroidetes*, *Actinobacteria*, and *Proteobacteria* (Moens & Veldhoen, 2002).

2.4.1.2.1 The intestinal mucosal immune system

The mucosal immune system has inductive and effector sites. The inductive sites are made up of lymphoid tissue and mucosa-draining lymph nodes, while effector sites refer to the lamina propria, stroma of exocrine glands, and the surface epithelia. The first line of defense in the intestines involves preventing the microbes in the gut lumen from penetrating the intestinal tissue and reaching the opposite surface. This is accomplished by the mucus layer, IgA gradients, and AMPs (Moens & Veldhoen, 2002; Liu *et al*., 2020). The mucus layer is produced from the polymerization of mucins or MUC2 secreted by goblet cells. In the large intestine, where there are a large number of bacteria, there are much more goblet cells than in the small intestine. These generate a very thick mucus that consists of two layers, the inner, firm one and an outer, loose layer. The inner layer is anchored to the intestinal epithelia and prevents microbial invasion so that it is sterile. This is made possible by the presence of IgA and defensins. The outer layer is inhabited by commensal bacteria that consume the nutrients in the mucin. The inner mucus layer is converted into the outer mucus layer by the proteolytic processing of polymerized mucin by the host or bacteria (Silva *et al.*, 2016; Okumura & Takeda, 2017).

The monolayer of epithelial cells forms the second line of defense (see Figure 2.32). It is made up of four types of cells, namely absorptive enterocytes, mucus-producing goblet cells, hormone-producing enteroendocrine cells, and AMP-producing Paneth cells (Moens & Veldhoen, 2002). The intestinal epithelial cells make up a physical barrier, which includes the glycocalyx on microvilli of enterocytes and the cell junctions that tightly connect the epithelial cells. Even if bacteria can penetrate the inner mucus layer, the epithelium prevents both pathogenic and beneficial microbes from invading the intestinal mucosa. The defensin family of AMPs and regenerating islet-derived 3 (Reg3) protein family produced by the Paneth cells make up the chemical barrier that keeps the intestinal microorganisms separated from the intestinal tissue. Defensins have antimicrobial activity against Gram-positive and Gram-negative bacteria, while the Reg3 family of proteins acts against Gram-positive bacteria. The epithelial cells also modulate host immune responses by secreting cytokines and chemokines(Okumura & Takeda, 2017).

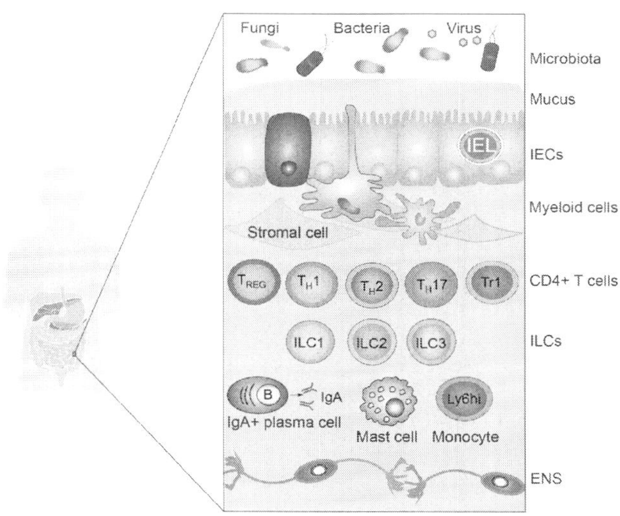

Figure 2.32. A mucosal layer and a single layer of intestinal epithelial cells (IECs) separate the gut microbiota from the underlying tissues. There are in-traepithelial lymphocytes (IELs) between the epithelial cells that maintain the mucosal barrier and provide protection from pathogens. Internal to the epithe-

lium is the lamina propria, which is a connective tissue composed of stromal cells, blood vessels, and nerves. It is also made up of immune cells, namely macrophages (blue myeloid cells) and dendritic cells (green myeloid cells). Other innate immune cells in this tissue layer include mast cells, monocytes, neutrophils, eosinophils, T cells (i.e., subsets of CD4+ T cells: Treg, Tr1, Th1, Th2, and Th17), IgA-producing B cells, and innate lymphoid cells (i.e., ILC1, ILC2, and ILC3). (Source: Sara M. Parigi *et al*., Creative Commons License)

On the other hand, the intestinal mucosa provides an ideal environment to promote microbial colonization, and the integrity of this epithelial layer allows the lumen bacteria to communicate with the gut immune system. Thus, the surface epithelium has a critical role in maintaining tolerance toward the gut microbiota since unnecessary immune responses to these microorganisms can lead to intestinal inflammation. This layer also alters its energy source to favor some microbes over others. For instance, fucosylated glycan can be utilized by certain bacterial species as their energy source to outcompete others. Such species recruit accessory species that are of benefit to them and induce morphological changes to the intestinal barrier like activating antimicrobial factors against their competitors. All these allow the colonists to take full control of their environment (Moens & Veldhoen, 2002; Silva *et al*., 2016).

Apart from influencing host immunity, the intestinal microbiota digests dietary fibers into short-chain fatty acids that can be used by the host as an energy source and metabolizes bile acids. Moreover, it synthesizes vitamin B and vitamin K (Okumura & Takeda, 2017).

The third line of defense involves intraepithelial lymphocytes (IELs), which is made up of a heterogeneous population of mostly unconventional T cells that populate the epithelium before birth (see Figure 2.32). Similar to conventional T cells, IELs also express antigen receptors, but these are not as diverse as T cell receptors (TCRs). Unlike TCRs, IEL receptors maintain an active state so that they do not require a priming step before being fully activated. They immediately release cytokines upon activation (Moens & Veldhoen, 2002).

The lamina propria, which lies beneath the epithelium and is composed of connective tissue, blood vessels, enteric nervous system, and immune cells, is the final line of defense (see Figure 2.32). There are also macrophages and DCs adjacent to the epithelial layer, and they detect luminal pathogens as well as initiate innate and adaptive immune responses. Other immune cells in the lamina propria are mast cells, monocytes, neutrophils, eosinophils, and lymphoid cells. Different subsets of CD4 T cells and B cells also accumulate in the lamina propria after being primed in the draining lymphoid tissues. The gastrointestinal mucosa is also enriched with innate lymphoid cells 1(ILC1), ILC2, and ILC3 (Moens & Veldhoen, 2002; Parigi *et al*., 2015).

The gut-associated lymphoid tissues (GALT) make up the inductive sites of the gut immune system. It is made up of mesenteric lymph nodes, isolated lymphoid follicles, and Peyer's patches (Parigi *et al*., 2015). Along with the nasopharyngeal-associated lymphoid tissues (NALT), it is a part of the mucosa-associated lymphoid tissue (MALT) network that provides a continuous source of memory B and T cells that migrate to the effector sites. The MALT is covered by microfold (M) cells and epithelial cells with lymphoid cells underneath that initiate mucosal immune responses (see Figure 2.33). The MALT covering has no goblet cells so that it is not covered by mucus. This enables the M cells to take up antigens from the lumen and transport them to the underlying DCs. The DCs, then, carry the antigens to the Peyer's patch or, through the draining lymphatics, to the mesenteric lymph nodes, where mucosal T and B cell responses are initiated. Some DCs produce retinoic acid that enhances the expression of mucosal homing receptors, such as $\alpha_4\beta_7$ and CCR9, on activated T cells. These activated T cells will then be guided to migrate through the lymphatic system, the bloodstream, and into the lamina propria where it triggers adaptive immune responses (McGhee & Fujihashi, 2012).

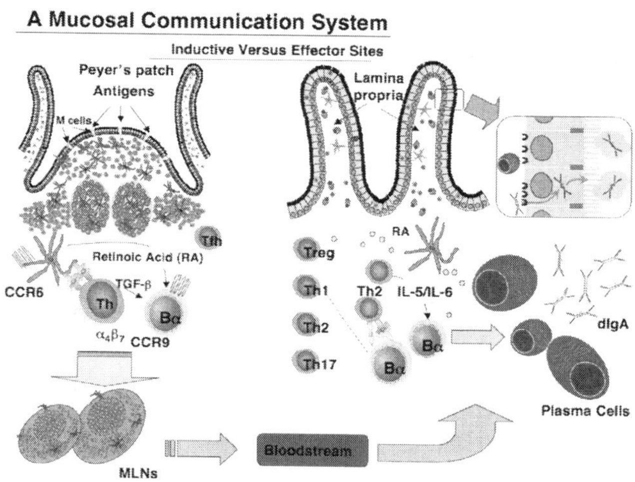

Figure 2.33. The gut-associated lymphoid tissues (GALT) include Peyer's patches and mesenteric lymph nodes (MLN). Microfold (M) cells in the epi-

thelial layer covering GALT sample antigens, which are then ingested by dendritic cells (DCs) and subsequently trigger specific T and B cell responses in the Peyer's patches and MLNs. Retinoic acid produced by DCs enhances the expression of homing receptors (e.g., $\alpha_4\beta_7$ and CCR9) on lymphocytes, which are then guided to enter major effector sites like the lamina propria. (Source: Jerry R. McGhee & Kohtaro Fujihashi, Creative Commons License)

Most PRRs are present in the intestine, and these can distinguish luminal antigens that are required for epithelial function and wound repair from those that invade the sterile tissues and pose threat to the host's well-being. This is possible partly because of the polarized structure of the barrier sites or the alternative signaling pathway. Activation of the inflammasome is also involved in maintaining intestinal homeostasis (Moens & Veldhoen, 2002). The commensal microbes influence the epithelial cells to preserve microbial homeostasis. They also secrete modulatory proteins that regulate the interaction between epithelial cells and immune cells to prevent the invasion of pathogens. Specifically, these proteins alter epithelial cell gene expression and influence MHC II expression, thereby regulating antigen presentation to immune cells. The *Clostridia* family downregulates the expression of a key enzyme for vitamin A metabolism regulating retinoic acid concentration in gut epithelial cells. This controls the stimulation of T cells and group 3 innate lymphoid cells to produce IL22. Thus, these commensal microbes modulate the IL22 immune response to prevent microbial dysbiosis and pathogen invasion (Larsen *et al*., 2020).

The epithelial cells of the gastrointestinal tract are quickly replaced via a process that occurs at the bottom of the small intestine and colon crypts where intestinal cells proliferate. The maintenance of the epithelial layer is greatly influenced by the gut microbiota, which regulates pH, affects cellular metabolism, and sets up immunity. For example, they enhance the degradation of complex carbohydrates into monosaccharides and short-chain fatty acids, which regulate the growth and differentiation of epithelial cells (Moens & Veldhoen, 2002; Larsen *et al*., 2020).*E. coli*, for instance, stimulates the production of mucus and establishment of protective mechanisms (Moens & Veldhoen, 2002; Larsen *et al*., 2020). Another example is *Enterococcus faecium* that stimulates the release of secreted antigen A (SagA), which increases the expression of barrier effectors and AMPs in the epithelial cells (Larsen *et al*., 2020).

2.4.1.2.2 Mechanism of gut immunity

Once the presence of pathogens is detected, epithelial cells upregulate the synthesis of defensins, which provide immediate protection, and alarmins, which alert the adjacent epithelial cells and the immune system (see Figure 2.34). Paneth cells produce α-defensins and most intestinal epithelial cells produce β-defensins. The presence of the pathogenic *Salmonella enterica* triggers the intestinal epithelial cells to express PRRs, such as TLRs and NOD (i.e., NLRP6). TLR4/Myd88 and NOD2 signaling regulate the production of AMPs by Paneth cells. These AMPs are no longer restricted to the enterocytes but are expressed across the entire epithelium. Also, the activation of TLRs induces a cascade that leads to NFκB activation and the subsequent production of pro-inflammatory cytokines like IL12 and IL6. Microbial and viral DNA in the cytosol can also activate the AIM2 inflammasome leading to the activation of caspase 1 and the release of pro-inflammatory cytokines IL1β and IL18. Another sensor of cytosolic DNA is **cyclic** GMP-AMP synthase (cGAS). When activated, it produces cGAMP that activates the stimulator of interferon response genes (STING) and establishes a strong antiviral response. The inflammasome and cGAS pathway may crosstalk with each other since IL1β can signal to neighboring epithelia to release mitochondrial DNA into the cytosol, which then activates STING (Larsen *et al.*, 2020). Activation of the TLR signaling pathway by the recognition of bacteria triggers the intestinal epithelial cells to produce a proliferation-inducing ligand (APRIL). APRIL drives IgM to IgA class switching in B cells found in the lamina propria (Okumura & Takeda, 2017).

The recognition of microbes or their metabolites (e.g., short-chain fatty acids) or the presence of Th2 cytokines (e.g., IL5 and IL13) can stimulate the goblet cells to secrete mucus. NLRP6 activation leading to the stimulation of inflammasome can also promote goblet cell autophagy resulting in the release of mucus granules. Meanwhile, Gram-negative bacteria like *Escherichia* and *Proteus* stimulate TLR5/Myd88 signaling in the intestinal epithelium so that IL8 is produced. This recruits neutrophils to the lamina propria (Okumura & Takeda, 2017).

During parasitic infection, chemosensing epithelial cells called Tuft cells detect microbial dysbiosis. They have receptors that recognize succinate released by certain microbes. When these microbes expand, the Tuft cells expand correspondingly. This results in increased levels of Tuft cell-derived IL25, which then activates innate lymphoid cell type 2 (ILC2) and Th2 cell immune response. ILC2s produce IL13, which influences intestinal stem cells to favor differentiation into Tuft and goblet cells. Damaged intestinal cells also trigger Th2 response by releasing IL33 that activates ILC2 to secrete IL5 and IL3. Thymic stromal lymphopoietin (TSLP) is another cytokine produced by intestinal epithelial cells that induce Th2 cell immune responses in response to parasites (Larsen *et al.*, 2020; Okumura & Takeda, 2017).

In the case of inflammatory bowel disease (IBD), a group of diseases characterized by chronic inflammation of the digestive tract, the affected tissues activate macrophages, which can eradicate certain pathogens using proteases and free radicals (see Figure 2.34). Then the major histocompatibility complex that is expressed on the cell membrane presents the antigens to T cells. During an IBD acute phase, macrophages in the intestinal mucosa significantly increase in number and they express a large number of cytokines and costimulatory molecules involved in the inflammatory response. The TLRs in DCs are also activated, which leads to the upregulation of pro-inflammatory cytokines, such as IL6 and IL12. DCs also transport antigens to GALT, where the naïve T cells are activated into effector Th1, Th2, and Th17 cells as well as where the B cells mature to produce immunoglobulins, especially IgG, IgM, and IgA. Th1 cells produce a variety of pro-inflammatory cytokines, such as IL1, IL2, IL6, IL8, IL12, TNFα, and IFNγ. IL6 promotes the proliferation of intestinal epithelial cells and tissue healing. In contrast, IFNγ inhibits epithelial cell proliferation and activates more macrophages. Still, these macrophages are recruited to the mucosal wound site to promote epithelial regeneration. Th2 cells secrete IL4, IL5, IL9, and IL13, which regulate the differentiation and activation of B cells. IL5 and IL13 also induce alternative macrophage activation that contributes to epithelial cell proliferation to promote wound healing. However, upregulation of IL13 promotes the apoptosis of intestinal epithelial cells resulting in the disruption of the mucosal layer. Th17cell activation leads to the production of IL17, which also promotes the inflammatory response, and IL22. IL17 and IL22 upregulate the expression of AMPs and the Reg3 family of proteins. IL22 also induces the synthesis of fucosyltransferase 2 (Fut2), which is involved in the glycosylation of intestinal epithelial cell-surface proteins. Fucosylation of membrane proteins in intestinal epithelial cells is important for protection against pathogens. On the other hand, Treg cells exert an anti-inflammatory effect by inhibiting other Th cell subtypes viaIL10 and TGFβ (Silva *et al.*, 2016; Okumura & Takeda, 2017)

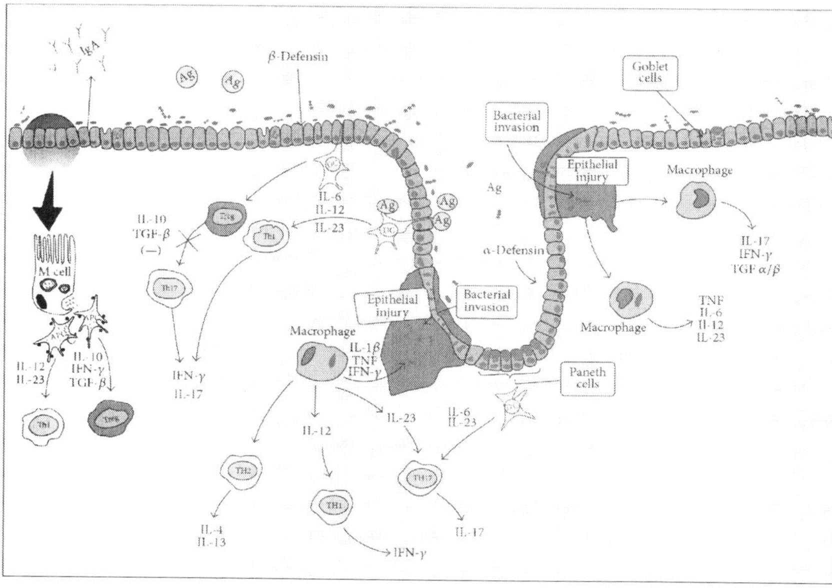

Figure 2.34. When bacterial invasion or epithelial injury is detected, such as in the case of inflammatory bowel disease (IBD), epithelial cells increase the production of α-defensins and β-defensins. The number of macrophages also dramatically increases, and these macrophages express costimulatory molecules of the inflammatory process. The Toll-like receptors (TLRs) on dendritic cells (DCs) are also activated so that there is increased expression of pro-inflammatory cytokines. DCs are also involved in the activation of naïve T cells into the pro-inflammatory Th1, Th2, and Th17 cells as well as the anti-inflammatory Treg cells. Moreover, they are involved in the maturation of B cells in order to produce antibodies such as IgA. (Source: Francesa A.R. Silva *et al.*, Creative Commons License)

S. enterica infection can influence the differentiation of intestinal stem cells so that they are more likely to specialize to become enterocytes and Paneth cells. On the other hand, other pathogens like *Clostridium difficile* can trigger infected epithelial cells to undergo apoptosis in order to prevent the spread of infection (Larsen *et al.*, 2020).

2.4.1.3. Immunity in the respiratory epithelium

Humans inhale 10,000 liters of air every day. Inhaled air contains life-sustaining oxygen, but it also has microbes, toxins, and pollutants, which present danger to the respiratory organs. The alveoli of the lungs are lined by epithelial cells, which are intimately associated with pulmonary capillaries. This structural interaction allows for the exchange of gases between the air and systemic circulation. Hence, if the air that reaches the alveoli is not filtered and sterilized, pathogens can invade the circulatory system. This is prevented from occurring by multi-layered chemical and physical defense mechanisms in the epithelium starting from the nasal region to the airways. The nasal epithelium produces large amounts of AMPs. For the airways, protection is accomplished by their branching structure, layers of mucus, tight adhesions between the epithelial cells and the underlying stroma, and mucociliary clearance via fluid and antimicrobial compounds. As for the alveoli, surfactants play a critical role in maintaining their structural stability and in mounting antimicrobial responses (Whitsett &Alenghat, 2015).

Exposure of the generally sterile lungs to pathogens normally causes an inflammatory response (Parker & Prince, 2011). Interestingly, the lungs actually host certain commensal bacteria. Therefore, the mucociliary barrier as well as the innate and adaptive immune responses in the respiratory tract aim to eliminate pathogens and maintain tissue-microbiome homeostasis (Whitsett & Alenghat, 2015).

2.4.1.3.1 Immune responses in the airways

Air that is taken is initially conducted through cartilaginous airways composed of the trachea that branches out into two bronchi, which further branch out into numerous bronchioles that extend deep into the lung parenchyma. The airways are lined with pseudostratified epithelium made up mainly of ciliated cells. The cilia on these cells move in a coordinated and directional manner, which enables the transport of particles and pathogens from the lungs (Whitsett & Alenghat, 2015).

The airway epithelium forms a tightly linked barrier separating the inhaled air in the lumen from the underlying tissues. The impermeability of this barrier is due to the adherens junctions, apical tight junctions, and hemidesomosomes, which make up the luminal junctional complex (see Figure 2.35). This complex is disturbed by inhaled substances like pollutants, microbes, or allergens. The airway epithelial cells closely interact with the lung microbiome to maintain the epithelial barrier and homeostasis (Hiemstra *et al.*, 2015). The adherens junctions appear as spots, known as adhesion plaques, or as bands encircling the cell, which are called zonula adherens. Adhesion plaques attach cells to the extracellular matrix, while zonula adherens interconnect the actin filaments of adherent cells.

The main regulator of paracellular permeability is the zonula occludens, a multiprotein junctional complex formed by tight junctions. In the airway of healthy individuals, the tight junctions of the zonula occludens and the apical junctions of the zonula adherens form dense protein networks that interconnect the basolateral sides of epithelial cells so that they prevent the paracellular passage of all molecules, like water, ions, proteins, as well as pathogens and other particles (Frey *et al*., 2020). Thus, disruption of the barrier and injury to the lungs is prevented (Leiva-Juárez *et al*., 2018). Hemidesmosomes, on the other hand, form adhesive bonds between the cytoskeleton of epithelial cells and the lamina lucida, which is a part of the lamina propria (Frey *et al*., 2020). The airway epithelium is also made up of secretory cells, such as serous, club, neuroendocrine, and goblet cells, as well as regenerative basal cells. There are also submucosal glands in the airways, and these comprise of terminal serous and mucous cells as well as collecting duct cells. Different parts of the airways exhibit varying proportions of these cell types. For example, the most distal bronchioles usually do not possess goblet cells (Whitsett & Alenghat, 2015; Hiemstra *et al*., 2015).

Secretory cells produce mucins, antimicrobial compounds, such as human-β-defensins(hBD1, hBD2, hBD3, and hBD4), lysozyme, lactoferrin, cathelicidin LL37, secretory leukocyte proteinase inhibitor (SLPI), and other fluids like surfactant protein A and surfactant protein D (Whitsett & Alenghat, 2015; Hiemstra *et al*., 2015). AMPs are also produced by phagocytic cells in the airways, whereas neutrophils in the lungs are also a major source of the cathelicidin LL37. Some AMPs are produced constitutively, such as hBD1, while hBD2, hBD3, and hBD4, which are induced by bacterial, fungal, and viral products as well as inflammatory mediators. These kill a wide range of pathogens, which include *E. coli*, *Pseudomonas*, and *H. influenzae*. LL37 specifically acts against *P. aeruginosa* and *S. aureus* possibly by electrostatically disrupting their membranes. It also enhances bacterial phagocytosis in macrophages. SLP1 exhibits antimicrobial activity against certain bacteria (*P. aeruginosa*, *Staphylococcus aureus*, and *E. coli*), fungi (*C. albicans* and *Aspergillus fumigatus*), and the human immunodeficiency virus(HIV). Another antiprotease is elafin that directly acts against *P. aeruginosa* and *S. aureus*. An AMP produced in the upper airways is the short palate lung and nasal epithelium clone 1 (SPLUNC1), which binds bacterial lipopolysaccharide and is shown to protect mice from *P. aeuroginosa* and *Mycoplasma* (Sharma *et al*., 2020). It may also function as a surfactant protein and biofilm formation inhibitor (Leiva-Juárez *et al*., 2018).

AMPs also have regulatory functions. For instance, α-defensins produced by neutrophils induce the proliferation of airway epithelial cells and thus, facilitate the repair of epithelial wounds. Defensins and LL37 are also induced by infections to act as chemoattractants for neutrophils, DCs, T cells, macrophages, and monocytes. LL37 also plays a role in angiogenesis, healing of epithelial wounds, and lung cancer growth. Antiproteases modulate macrophages and endothelial cells resulting in decreased IκBα degradation and increased synthesis of TGFβ and IL10 (Leiva-Juárez *et al.*, 2018; Hiemstra *et al.*, 2015).

A high concentration of AMPs may be toxic to cells. For instance, α-defensins cause lysis of lung epithelial cells and induce IL8 production in these cells so that high concentrations lead to uncontrolled inflammation in certain diseases like cystic fibrosis and chronic bronchitis (Bals & Hiemstra, 2004).

Lysozyme and lactoferrin are much larger than AMPs and are abundant in the airways. Lysozyme disrupts the β, 1→4 glycosidic bonds between N-acetylglucosamine and N-acetylmuramic acid in the peptidoglycan of Gram-positive bacteria. On the other hand, lactoferrin chelates iron, an essential element for the growth of replicating organisms, and other components away from the bacterial membrane of Gram-negative bacteria so that their membrane is disrupted and the peptidoglycan is exposed and targeted by lysozyme (Parker & Prince, 2011). Another secreted protein, lipocalin 2, is similar to lactoferrin in its ability to sequester iron. It responds against *K. pneumoniae* and *E. coli* (Leiva-Juárez *et al.*, 2018). All these substances kill or inhibit the propagation of pathogens until they are eliminated by the mucociliary apparatus, recruited phagocytes, and/or adaptive immune responses (Bals & Hiemstra, 2004).

Mucins are large glycoproteins that are rich in repeated threonine domains with complex O-linked polysaccharides. They create a barrier and "rafts" on the luminal surfaces of airways even down to the alveoli, Such barriers bind to and transport pathogens so that they do not invade the epithelial cells underneath. The mucins on the epithelial surface can then be degraded by pathogen- or host-associated proteases and the entrapped microbes are subsequently released to the mucociliary barrier for removal (Whitsett & Alenghat, 2015). Mucins are negatively-charged and interact electrostatically with other molecules, many of which have antimicrobial activity, such as IgA, AMPs, lysozyme, and collectins (Hiemstra *et al.*, 2015).

The static mucin barrier along with glycolipids form the glycocalyx or periciliary layer (PCL), which can vertically stick out of the apical surface of the epithelium, reaching a height of up to 1,500 nm in some areas (see Figure 2.35). It stores water to control mucus hydration and protects from the compression of the overlying mucus to allow the persistent beating of the cilia during mucociliary transport. (Frey *et al.*, 2020). Particles of about 100 nm and above, such as influenza and parainfluenza virus, adenovirus or coronavirus, are efficiently blocked by the PCL. In contrast, very small viruses with sizes ranging from 20-30 nm, like bocavirus and human rhinovirus (HRV), can penetrate the PCL and infect airway epithelial cells if their receptors are present on the apical side.

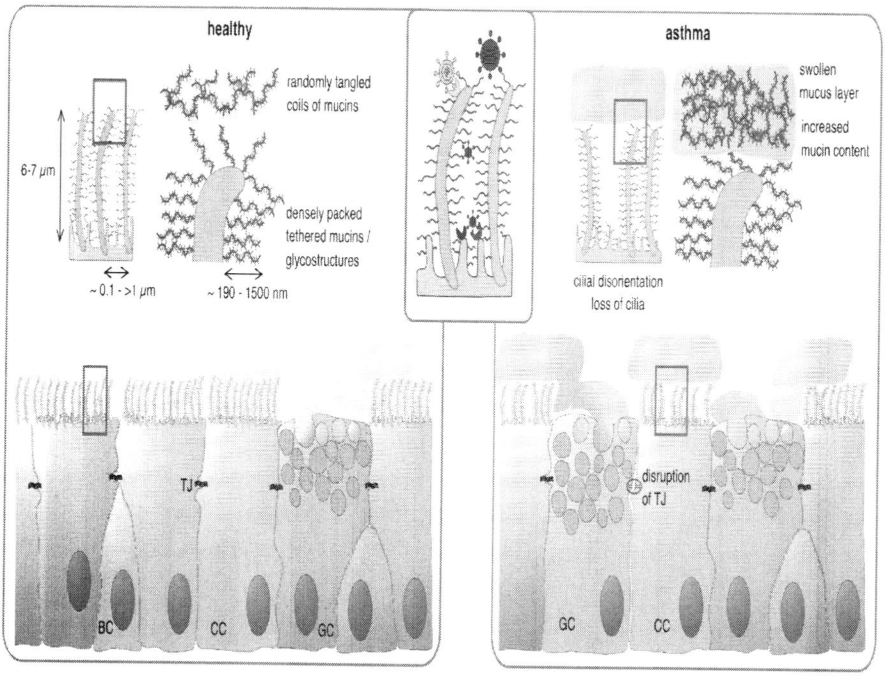

Figure 2.35. The respiratory tract is protected by physical barriers like the epithelial layer made up of ciliated cells (CC) that are firmly connected by junctional complexes, one component of which is the tight junction (TJ). These complexes seal the paracellular spaces so that passage of harmful pathogens and substances is prevented. Another barrier is formed by mucins (green), secreted by goblet cells (GC), and glycolipids that form a layer of glycocalyx over the cilia, which can protrude up to 1,500 nm from the apical surface of the epithelium. This restricts the access of substances from the lumen to the apical

cell surface so that large-sized pathogens can be cut off from their receptor (inset). Mucins with absorbed water also form a viscous gel or mucus that float on the surface barriers and trap microbes and particulate matters so that these are transported upward through the airway lumen via the coordinated beating of the cilia. (Source: Andreas Frey *et al.*, Creative Commons License)

Respiratory infection or exposure to irritants is characterized by increased synthesis and release of mucus into the airways. The mucins that are secreted into the airways combine with water, salts, lipids, and proteins to form a sticky mucus gel that disrupts the aggregation of pathogens, dust, and other harmful objects by binding to them and then preventing them from attaching to the epithelial surface (see Figure 2.35). This facilitates mucociliary clearance of microbes from the airways (Whitsett & Alenghat, 2015; Frey *et al.*, 2020). Mucus is moved along the airway epithelium toward the larynx via the constant and coordinated beating of the cilia. This mucociliary transport or escalator is a critical mechanism for the clearance of inhaled or aspirated microbes or particles (Hiemstra *et al.*, 2015). It enables the expulsion of particles and microbes from the lungs to the oral cavity within minutes (Leiva-Juárez *et al.*, 2018). Then, the mucus can be expectorated. Only nanoscalar or smaller molecules, which include gases, water, salts, and nutrients, can reach the epithelial cells since they can diffuse faster toward the epithelial surface relative to the movement of mucus (Frey *et al.*,2020). The frequency of the ciliary beat depends on the presence of mechanical stress as well as neurochemical and inflammatory signals. Such signals trigger the intracellular movement of calcium and intercellular transport of inositol triphosphate. The gap junctions allow rapid exchange of responses to stimuli among ciliated cells. For instance, *Pseudomonas aeruginosa* activates TLR2, which then drives the gap junction protein connexin Cnx43 to mediate calcium-dependent signaling. This contributes to NF-κB activation and secretion of cytokines and chemokines into the submucosa, which attracts neutrophils and T cells to the site of pulmonary infection (Whitsett & Alenghat, 2015).

The mucus layer generally prevents microbes from directly coming in contact with the epithelium. Nevertheless, PAMPs that are shed by these pathogens can permeate this layer and be recognized by PRRs, such as TLRs, RIG, CLRs, and NLRs, that are expressed in the epithelial cells (Parker & Prince, 2011; Whitsett & Alenghat, 2015). TLR4 is activated by lipopolysaccharide, respiratory syncytial virus, cigarette smoke, and inflammatory cytokine. Both TLR2 and TLR4 are activated by *Klebsiella pneumoniae*. TLR4 is also activated by house dust mites. *Haemophilus influenzae* lipoprotein or rhinovirus triggers TLR2- and TLR3-mediated signaling pathways as well as the expression of MUC5AC in conducting airways. *P. aeruginosa* flagella are recognized by TLR2 and TLR5. TLR activation also leads to the synthesis and release of AMPs, such as hBD2and hBD4, into the airway lumen. On the other hand, the RIG-I and TLR3-mediated signaling pathway is triggered by respiratory viruses. The NLRs NOD1 and NOD2 in airway epithelial cells are involved in clearing bacterial pathogens. *S. pneumoniae* produces a pore-forming toxin **pneumolysin** that facilitates the entry of peptidoglycan from *H. influenzae* to activate NOD1. NOD1 activation was also observed to reduce airway hyperresponsiveness and allergen-specific T cell proliferation during allergen-induced lung inflammation. There are also NLR inflammasomes in the airways, such as NLRC4 that recognizes the cytosolic flagellin of bacteria like *P. aeruginosa*, *K. pneumoniae*, and *Legionella pneumophila* and NLRP3 that senses bacterial peptidoglycan, asbestos, and uric acid from lung injury.NLRP3 can also detect potassium efflux, excess ROS production, and mitochondrial dysfunction. Dectin-1, a CLR found on bronchial epithelial cells, mediates the recognition of β-glucan motifs in the fungus *A. fumigatus* and the house dust mite. Another type of CLR, proteinase-activated receptor 2 (PAR-2), responds to nonfungal allergens like Derp1 and cockroach allergen (Whitsett & Alenghat, 2015; Parker & Prince, 2011; Hartl *et al.*, 2018; Leiva-Juárez *et al.*, 2018). Detection of a number of viral PAMPs by TLR3, TLR4, TLR7, TLR8, TLR9, NOD, and even RNA polymerase III, triggers the production of type I and type III IFNs. Therefore, if the epithelial defense fails to limit the invading pathogens, PAMP-PRR interaction triggers signaling pathways that ultimately recruit various immune cells to help eliminate the pathogens. The innate immune response triggered by the PAMP-PRR interaction efficiently clears the pathogens so that sterility of the lower airways is maintained (Hiemstra *et al.*, 2015; Parker & Prince, 2011).

The epithelial cells of the airways produce a large amount of ROS via the activity of NADPH oxidases, particularly dual oxidases (DUOX). Type II alveolar epithelial cells also express DUOX. ROS production by DUOX1 is enhanced by IL4 and IL13, while IFNγ increases DUOX2 expression. ROS contributes to the antimicrobial activity of the epithelial cells (Leiva-Juárez *et al.*, 2018; Hiemstra *et al.*, 2015). It was observed to increase the expression of RIG-I and MDA-5 in nasal epithelial cells that were infected with influenza A virus (Kim *et al.*, 2015). In addition, the activity of nitric oxide synthase (NOS) in airway epithelial cells produces reactive nitrogen species (RNS), such as nitric oxide (NO), which plays a role in immune regulation and host defense (Hiemstra *et al*., 2015). An interesting finding is that *P. aeruginosa* activates the bitter taste receptor T2R38, leading to increased NO production and enhanced mucociliary clearance (Hartl *et al*., 2018). Lymphocytes also reside in the airway epithelium, specifically over the epithelial membrane and between the epithelial cells. The bronchial walls also have lymphoid tissues, comprising solitary lymphoid follicles or their aggregates that resemble the Peyer's patches of the intestine. This bronchus-associated lymphoid tissue (BALT) is analogous to the GALT of the intestine (see Figure 2.36). BALT is present in the lungs of kids and adolescents, whereas in adults it is present only during chronic inflammatory disease, and so it is called inducible BALT (iBALT). Its major immune cell population is made up of B cells that generate IgA, which protects against viruses, bacteria, and even allergens. It is also composed of T cell zones made up of T cells and DCs. BALTs also have the high endothelial venule (HEV) that transports lymphocytes and antigens to and from the bloodstream (Kumar, 2020).

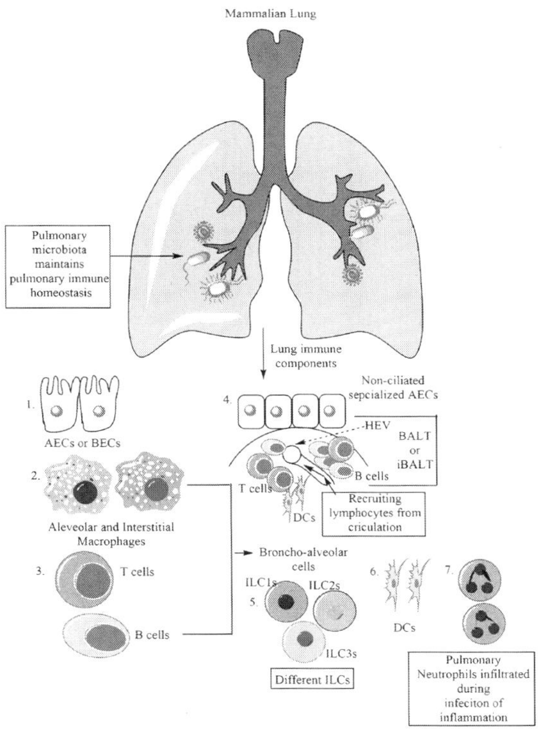

Figure 2.36. The respiratory epithelium made up of bronchial and alveolar epithelial cells (BECs and AECs) provides a protective barrier against pathogens

and particulates in inhaled air. It is complemented by immune cells present in the respiratory system, namely macrophages, dendritic cells (DCs), subsets of innate lymphoid cells (ILC1, ILC2, and ILC3), and neutrophils. Lungs also have bronchus-associated lymphoid tissue (BALT), which contains B cells, T cells, and DCs, and is induced to become iBALT in response to infection. (Source: Vijay Kumar, Creative Commons License)

2.4.1.3.2. Immune responses in the alveoli

The alveoli are also lined with epithelial cells, which are of two types. An estimated 90% of the alveolar surface in the adult lungs are made up of type I epithelial cells or pneumocytes which are squamous and closely interact with endothelial cells of pulmonary capillaries so that they are optimized for gas exchange. The other cells in the alveolar lining are cuboidal type II epithelial cells. These contain a lot of lamellar bodies that are rich in lipids and microvilli on their apical side. They are essential in the production of surfactant lipids and surface tension-reducing proteins. Moreover, these also function as self-renewing cells and precursors of type I cells during alveolar repair (Whitsett & Alenghat, 2015; Leiva-Juárez *et al*., 2018).

Despite being made up of a thin monolayer of cells, the alveolar epithelium creates multiple barriers comprised of their secretions, surface glycocalyces, membranes, and junctional proteins between cells. The junctional barrier is made up of claudins, connexins, paranexins, adhesions, and zonula occludins that interact with actin and provide structural integrity to the respiratory epithelium. Claudins, in particular, are highly expressed in alveolar cells and they create tightly interlocked structures that link these cells. When junctional complexes are compromised, this increases epithelial permeability and inflammation in both the airways and alveoli leading to certain respiratory illnesses (Whitsett & Alenghat, 2015).

Pulmonary surfactant reduces the surface tension at the gas-liquid interface at the alveolar surface. This is critical for lung structure and function since surface tension pulls water molecules closer together causing the alveoli to be pulled inwards as well. Without surfactants, the alveoli would collapse. Pulmonary surfactant is made up of 90% lipids and 10% proteins (Seadler *et al*., 2021). It is important that immune-related proteins produced in the alveoli should not disrupt but should enhance the surface activities of lipids. The type II alveolar epithelial cells produce various surfactant proteins. A variety of PAMPs (e.g., lipopolysaccharide and bacterial lysates), glucocorticoids, hyperoxia, and IFNγ induce the lectin-like domains of

Surfactant protein A (SP-A) and SP-D. Afterward, alveolar macrophages are enhanced to opsonize and kill pathogens. Surfactants are also observed to disrupt the membrane permeability of certain Gram-negative bacteria. SP-B and SP-C produce small hydrophobic peptides required for surfactant lipid spreading and stability. SP-B also promotes the killing of bacteria at acidic pH within alveolar macrophages, whereas SP-C binds endotoxin (Leiva-Juárez *et al*., 2018; Whitsett & Alenghat, 2015).

Alveoli have resident macrophages and DCs that detect antigens and establish innate and adaptive immune responses (see Figure 2.36). These immune cells are observed to be physically linked via Cnx43 to type I and type II epithelial cells. This connection enables cell signaling that regulates local inflammatory responses to injury (Whitsett & Alenghat, 2015). Alveolar macrophages are the predominant phagocyte and APC in the human respiratory system. They are the most abundant cellular fraction (i.e., 90%) in bronchoalveolar lavage fluids under healthy conditions. They serve as airway scavengers ingesting pollutants, allergens, airborne microbes, other environmental particles, and even apoptotic cells. They also maintain surfactant homeostasis by engulfing and catabolizing surrounding surfactant proteins. On the other hand, they inhibit inflammation by producing anti-inflammatory cytokines upon sensing PAMPs and DAMPs (Hartl *et al*., 2018). Their anti-inflammatory activity is critical in maintaining the homeostasis of other alveolar cells, epithelial cells, DCs, and T cells. There are also macrophages located within the lung tissue parenchyma called the interstitial macrophages. These are important sources of cytokines under homeostatic conditions and upon exposure to environmental stimuli. They are not as phagocytic as alveolar macrophages, but bacterial lipopolysaccharide enhances their ability to engulf small particles. Lipopolysaccharides also stimulate their chemotaxis and their production of ROS. In terms of antigen presentation, they are superior relative to alveolar macrophages, and they drive Tcell and Treg cell differentiation. They are also involved in wound healing and tissue repair. Meanwhile, the primary role of DCs in the respiratory tract is to sample inhaled pathogens before migrating to lymph nodes where they present processed peptides to antigen-specific T cells (Ardain *et al*., 2019).

The remaining 10% of cells in the bronchoalveolar lavage fluids comprise lymphocytes (see Figure 2.35). These circulate, via the lymph, through the lungs and detect antigens inhaled into the lungs (Kumar, 2020). There are also resident ILCs in the lung, which have three subsets, i.e., ILC1, ILC2, and ILC3. They mediate immune responses against pathogens and parasites by producing cytokines and chemokines, and they also promote tissue repair and homeostasis following infections. NK cells are considered to be members of the ILC1 family, but they perform a distinct cytolytic function via their secretion of perforin and granzyme (Ardain *et al.*, 2019).

Type II alveolar cells express TLRs and are enriched in RAGE, a PRR that is activated by DAMP, such as high mobility group box 1 (HMGB1). HMGB1 is released during cell death and cytokine stimulation. It signals to nearby cells via TLR- and RAGE-dependent pathways influencing epithelial gene expression and cytokine production (Whitsett & Alenghat, 2015).

2.4.2 Physiological Defenses

As indicated by the previous discussion, the anatomical barriers also afford a certain level of physiological protection. Still, there are more defense mechanisms of the human body that depend on physiological factors including body temperature, pH, and chemical substances apart from the previously discussed AMPs.

2.4.2.1 Fever and immunity

The hallmark response to infection that is related to body temperature is **fever**. Febrile temperature has been one of the major signs of inflammation. The increase in core body temperature by 1°C to 4 °C during fever is correlated with survival and resolution of many infections. In particular, temperatures at 40-41 °C lead to more than 200-fold reduction in the replication rate of poliovirus in mammalian cells and to increased susceptibility of Gram-negative bacteria to serum-induced lysis (Evans *et al.*, 2015).

The innate immune system and neural network in the central and peripheral nervous systems are involved in the induction and maintenance of fever during infection. PAMP-PRR interaction (e.g., LPS-TLR4) initiates the signaling pathway that results in the expression of pyrogenic cytokines, namely IL1, IL6, and TNF, at the site of infection and in the brain. IL6, in particular, is critical for mediating fever, and it is synthesized by certain brain cells in response to local inflammatory stimuli. It acts in the brain to promote the synthesis of cyclooxygenase 2 (COX2), which oxidizes arachidonic acid to produce prostaglandin E2 (PGE2). IL6 also directly stimulates the synthesis of another brain-resident cytokine, receptor-activator of NF-κB ligand (RANKL), which interacts with its receptor RANK and is possibly involved in the COX2-PGE2 pathway (Evans *et al.*, 2015).

PGE2 is produced by endothelial cells in the brain blood vessels and by innate immune cells, like DCs and macrophages. It binds to EP3 prostaglandin receptors, which are expressed by thermoregulatory neurons in the median preoptic nucleus in the hypothalamus. In a rodent model of fever induction, the input signal from pyrogenic cytokines is integrated by PGE2, which is considered to be the major pyrogenic mediator of fever, withoutout signals involving neurotransmitters that raise core body temperature. Norepinephrine also increases body temperature by accelerating thermogenesis in brown adipose tissues and inducing vasoconstriction in the extremities to prevent passive heat loss. Moreover, the neurotransmitter acetylcholine stimulates muscles to convert chemical energy into thermal energy and increases overall metabolic rates (Evans *et al*., 2015).

Apart from initiating the fever response, pyrogenic cytokines also establish immunity within the infected tissues. Febrile temperatures stimulate almost every step involved in initiating innate and even adaptive immunity. Data from animal models reveal that fever-range temperatures enhance the respiratory burst that is usually associated with neutrophil activation. Thermal stress also increases neutrophil recruitment to local sites of infection and other distant tissues. Such heat-induced neutrophil recruitment depends on the enhanced production of IL17, IL1β, IL1α, IL8, and heat shock protein (HSP) (Evans *et al*., 2015).

Fever-range temperatures also increase the cytotoxic activity of NK cells as well as their recruitment to tumor sites. Heat-induced enhanced cytotoxicity is due to a variety of factors. One of which is the heat-induced upregulation of MHCI polypeptide-related sequence A (MICA), the ligand of NKG2D receptor, on tumor cells and clustering of NKG2D receptors on NK cell surface. Other factors include the decreased expression of MHCI in tumor cells, while HSP70 expression is increased (Evans *et al*., 2015). T cells are also induced by febrile temperatures to produce HSP90. These proteins bind the cytoplasmic tails of α4 integrins, which are cell adhesion molecules, and induce the binding of talin and kindlin-3, resulting in α4 integrin activation. Also, HSP90s simultaneously bind to the tails of two α4 integrinson the inner surface of the cell membrane causing them to dimerize and aggregate on the T cell membrane. Integrin clustering activates the FAK-RhoA GTPase signaling pathway, which promotes T cell adhesion and migration along the walls of HEVs. This facilitates T cell trafficking to draining lymph nodes and inflamed tissues, thus enhancing immune surveillance during infection (Lin *et al*., 2019).

Hyperthermia does not only induceHSP70 upregulation, but it also causes the release of HSP70 into the extracellular environment where it can function as DAMP to stimulate macrophages and DCs. Macrophages, in particular, are stimulated to exhibit sustained activity in response to lipopolysaccharide. As a result, these produce pro-inflammatory cytokines, such as TNF, IL1, and IL6. HSP70 also stimulates peritoneal macrophages to express nitric oxide and nitric oxide synthase (Evans *et al*., 2015).

DC stimulation by elevated temperature results in its enhanced phagocytic function, increased IFNα production in response to a viral infection, and heightened pathogen recognition due to upregulation of TLR2 and TLR4. Fever-range temperatures also increase the DC expression of MHCI and MHCII as well as co-stimulatory molecules like CD80 and CD86 and enhance DC secretion of IL12 and TNF. Moreover, elevated temperatures augment the migration of APCs, such as Langerhans cells in the skin, to the draining lymph nodes. These observations may explain why thermal stress enhances the Tcell stimulating and cross-presenting abilities of DCs (Evans *et al*., 2015).

2.4.3 Phagocytosis

Phagocytosis is a process in which the cell ingests particles larger than 0.5 μm in diameter. For unicellular organisms, this is a way to nourish themselves. In multicellular organisms, many cell types perform phagocytosis, but professional phagocytes are specialized to perform this process with high efficiency. These efficient phagocytes include macrophages, neutrophils, eosinophils, monocytes, mature DCs, and osteoclasts, and they are responsible for containing and removing microorganisms as well as presenting antigens to lymphocytes in order to trigger the adaptive immune system (Uribe-Querol & Rosales, 2020; Harris, 2021). In fact, phagocytosis of I-E MHC class II molecules from cellular fragments is thousands of times more efficient in producing MHCII peptide complexes than taking in preprocessed I-E peptide. In addition, resident lymphoid DCs phagocytose short-lived migratory DCs and then present the antigens that were internalized by these migratory DCs (Garrett & Mellman, 2001). The incomplete digestion of engulfed pathogens also results in antigens with a size suitable for binding with MHC and presentation to lymphoid cells (Horiguchi *et al*., 2018). On the other hand. non-professional phagocytes include fibroblasts, epithelial cells, and endothelial cells. They are not capable of engulfing microorganisms, but they eliminate dead cells as well as maintain homeostasis and promote wound healing (Uribe-Querol & Rosales, 2020).

The efficiency of phagocytes to engulf their target seem to be affected by their differentiation status. For instance, monocytes generally are less efficient in phagocytosis than neutrophils and macrophages, but their phagocytic capacity, their ability to phagocytize diverse targets, and their efficiency in complement-mediated phagocytosis are modified upon cell differentiation. Mature phagocytes are also better at engulfing IgG-opsonized particles than undifferentiated monocytes (Uribe-Querol & Rosales, 2020).

Phagocytosis involves several phases: detection of the particle to be ingested, activation of the internalization process, formation of a specialized vacuole called phagosome, and maturation of the phagosome and fusion with the lysosome, which contains degradative enzymes, to become phagolysosome (Uribe-Querol & Rosales, 2020).

2.4.3.1 Recognition of target particle

Circulating phagocytes recognize chemokines released from the site of infection causing them to migrate to that area (Horiguchi *et al.*, 2018). Then the receptors on the phagocyte plasma membrane identify and engage a variety of particles, such as microbes and apoptotic cells, and then initiate signaling pathways that induce phagocytosis. These receptors can either be the non-opsonic or opsonic type. Non-opsonic receptors identify PAMPs or DAMPs on the particle to be ingested. Table 1 presents some of these along with their PAMP ligands. These receptors include C-type lectins, such as Dectin-1, Dectin-2, Dectin-3 (or macrophage C-type lectin, MCL), MINCLE, DC-SIGN, and scavenger receptor A (SR-A). TLRs, on the other hand, do not function as phagocytic receptors but they can prepare the cell and interact with phagocytic receptors to enhance phagocytic efficiency. Dectin-1 detects yeast polysaccharide, Dectin-3 recognizes trehalose dimycolate present in the mycobacterial cell wall and α-mannans, and MINCLEalso binds trehalose dimycolate. DC-SIGN serves as a receptor for viruses, bacteria, and fungi by binding fucosylated glycans and mannose-rich glycans (Uribe-Querol & Rosales, 2020).

Table 2.1. Non-opsonic Phagocytic Receptors and their Microbial Ligands

Receptor	Ligands
Dectin-1	Fungal β-glucan Some yeast polysaccharides
MINCLE	Trehalose dimycolate of Mycobacteria

MCL	Trehalose dimycolate α-mannan
DC-SIGN	Fucosylated glycans Mannose-rich glycans
Mannose receptor (CD206)	Mannan
CD14	Lipopolysaccharide-binding protein
CD36	*Plasmodium falciparum*-infected erythrocytes
Scavenger receptor A	Lipopolysaccharide Lipoteichoic acid
MARCO	Bacteria

Source: Uribe-Querol & Rosales (2020), *Frontiers in Immunology* 11: 1066

Table 1 also includes receptors that are yet unclear in terms of their induction of phagocytosis. Still, these are involved in priming the phagocyte, bringing the particle close to the phagocyte, or mediating phagocytosis. These include CD14, CD36, scavenger receptor A (SR-A), and MARCO (Uribe-Querol & Rosales, 2020).

Apoptosis, or programmed cell death, is essential in order to maintain homeostasis in multicellular organisms. Millions of cells die by apoptosis daily, and these need to be cleared via phagocytosis in order to prevent the release of potentially toxic or immunogenic materials from the dying cells that may impair the structure and function of the surrounding tissues. Their phagocytosis is initiated by certain receptors that can detect molecules that occur only on the membrane of dying cells.

These molecules include lysophosphatidylcholine and phosphatidyl serine (PS). Table 2 lists the receptors for these DAMPs. PS is directly recognized by the receptors T cell immunoglobulin mucin 1 (TIM-1), TIM-4, stabilin-2, and brain-specific angiogenesis inhibitor 1 (BAI-1). Other receptors for apoptotic cells are the integrin αvβ5, CD36, and CD14. Another receptor, αvβ3, requires lactahedrin (also known as milk fat globule-EGF-factor 8 or MGF-E8) to bind first to PS so that it can strongly attach to the apoptotic cell (Uribe-Querol & Rosales, 2020).

Table 2.2. Phagocytic Receptors for Apoptotic Cells

Receptor	Ligands
TIM-1	Phosphatidylserine
TIM-4	Phosphatidylserine
Stabilin-2	Phosphatidylserine
BAI-1	Phosphatidylserine
Lactahedrin and αvβ3	Phosphatidylserine
CD14	Phosphatidylserine (?)
CD36	Oxidized phosphatidylserine
αvβ5	Apoptotic cells

Source: Uribe-Querol & Rosales (2020), *Frontiers in Immunology* 11: 1066

Some non-apoptotic cells, like activated B and T lymphocytes, express PS on their plasma membrane. These cells avoid being ingested because they also express molecular signals that enable them to evade phagocytosis. An example of such a signal is CD47, a ligand of the receptor signal regulatory protein α (SIRPα) that is found on phagocytes. Upon their interaction, SIRPα delivers a signal that inhibits actin assembly so that phagocytosis is prevented from taking place (Uribe-Querol & Rosales, 2020).

Opsonic receptors detect host-derived proteins, called opsonins, that are bound to certain particles in order to label them for phagocytosis. Opsonins are, in essence, acting as bridges between the phagocyte and the target particle. Some examples of opsonins are antibodies, fibronectin, complement, milk fat globulin, and mannose-binding lectin, while examples of opsonic receptors are FcRs and complement receptors (CRs) (Uribe-Querol & Rosales, 2020). Table 2.3 presents examples of opsonic receptors and the opsonins that they bind.

Table 2.3. Opsonic Receptors and their Ligands

Receptor	Ligand
FcγRI (CD64)	IgG1=IgG3>IgG4
FcγRIIa (CD32a)	IgG3≥ IgG1>IgG2>IgG4
FcγRIIIa (CD16a)	IgG1= IgG3>IgG2>IgG4
FcαRI (CD89)	IgA1, IgA2

CR1 (CD35)	Mannan-binding lectin, C1q, C4b, C3b
CR3 ($\alpha_M\beta2$, CD11b/CD18, Mac-1)	iC3b
CR4 ($\alpha v\beta2$, CD11c/CD18, gp190/95)	iC3b
$\alpha5\beta1$ (CD49e/CD29)	Fibronectin, vitronectin

Source: Uribe-Querol & Rosales (2020), *Frontiers in Immunology* 11: 1066

FcRs bind with varying affinities to the Fc region or the constant portion of different subclasses of IgG or IgA antibodies. When Fcγ receptors (FcγRs) bind the Fc part of IgG antibodies, they get clustered on the membrane of the cell and then trigger phagocytosis as well as other cellular responses. There are three types of FcγR, namely FcγRI (CD64), FcγRII (CD32), and FcγRIII (CD16) (see Figure 2.37). Their α-subunit specifically binds IgG. This subunit in FcγRI has a high affinity for IgG and has three Ig-like extracellular domains. In contrast, the α-subunit in FcγRII and FcγRIII exhibits low affinity to IgG and possesses two Ig-like domains. Therefore, they can only bind multimeric immune complexes. There are two isoforms of FcγRII, namely FcγRIIa that is found predominantly on phagocytic cells and FcγRIIb that is expressed mainly on B lymphocytes. Likewise, FcγRIII has two isoforms. FcγRIIIa is expressed on macrophages, NK cells, basophils, mast cells, and DCs, while FcγRIIIb is found only on neutrophils (Uribe-Querol & Rosales, 2020).

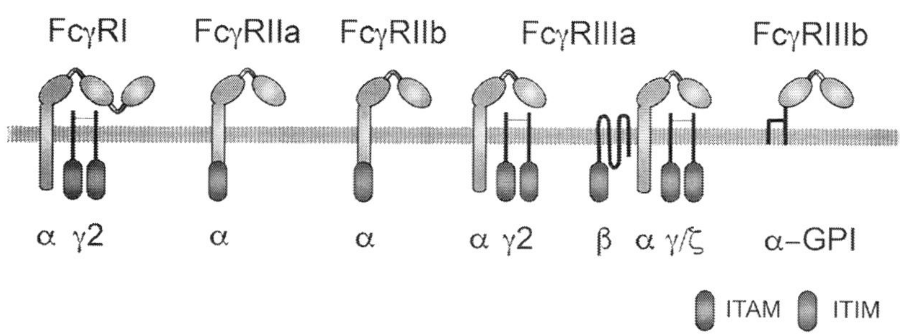

Figure 2.37. There are three groups of human Fcγ receptors, namely FcγRI, FcγRII, and FcγRIII. FcγRI has an α subunit with three Ig-like extracellular domains. Meanwhile, the α subunit of the other two types has two of these Ig-like domains. FcγRI and FcγRIIIa have an accessory dimer of γ chains which con-

tain an immunoreceptor tyrosine-based activation motif (ITAM). FcγRIIa has an ITAM in its α subunit. FcγRIIb, meanwhile, has an immunoreceptor tyrosine-based inhibition motif (ITIM) in its α subunit. Lastly, FcγRIIIb is bound to the plasma membrane via a glycosylphosphatidylinositol (GPI) anchor. (Source: Eileen Uribe-Querol & Carlos Rosales, Creative Commons License)

FcγRI and FcγRIIIa are expressed together with an accessory dimer of the Fc receptor gamma (FcRγ) chain with each chain having an ITAM sequence (see Figure 2.37).In contrast, FcγRIIa is not associated with FcRγ chains, but it has an ITAM motif in its cytoplasmic domain. When activatedFcγRs are clustered, the tyrosine residues in the ITAM sequence are phosphorylated. These residues are critical for receptor signaling. FcγRIIb, just like FcγRIIa, does not associate with FcRγ chains, but it has an ITIM motif in its cytoplasmic tail that is involved in negative signaling. Phosphorylated tyrosine residues within the ITIM recruit phosphatases that downregulate signals from ITAM-containing activated receptors. For this reason, FcγRIIb functions as a negative regulator of certain cellular processes including phagocytosis. FcγRIIIb has a conformation distinct from the other receptors. It is a glycosylphosphatidylinositol (GPI)-linked receptor that lacks a cytoplasmic tail and has no known associated subunits. It is also an activating receptor similar to those with ITAM (Uribe-Querol & Rosales, 2020).

Complement receptors (CRs) are another class of receptors that bind activated complement molecules that are deposited on microorganisms or cells. These are components of the complement system, which discriminates self from non-self as well as eliminates pathogens. Complement refers to a group of more than 30 proteins, produced in the liver and by tissue macrophages, that are activated in a sequential manner to enhance the immune responses of the body. There are three types of complement pathways. The classical and MB-lectin pathways are activated with the aid of PRRs, while the alternative pathway is activated spontaneously without requiring PRRs (Horiguchi *et al.*, 2018). The classical pathway is initiated by the direct binding of C1q, the first protein in the complement cascade, to the pathogen surface. It can also be activated by the binding of C1q to the antigen-antibody complexes during an adaptive immune response. The MB-pectin pathway is induced by the binding of the mannan-binding lectin, a serum protein, to mannose-containing carbohydrates on pathogens. Finally, the alternative pathway is initiated when a spontaneously hydrolyzed complement component, like C3, binds to the pathogen surface. These pathways generate C3 convertases that are covalently attached to the pathogen surface. They cleave C3 to generate C3b that acts as opsonin and C3a, a mediator of inflammation. A second molecule of C3b binds the C3 convertase to form the C5 convertase that produces C5a which mediates inflammation, and C5b that initiates the lytic pathway resulting in bacterial cell death (Janeway *et al*., 2001; Gani, 2021).

Complement receptors are subdivided into three groups. CR1 and CR2 form one group as they are formed by short consensus repeat (SCR) elements, CR3 and CR4 belong to the β2 integrin family, and CRIg belongs to the Ig superfamily (see Figure 2.38). The most efficient complement receptor in terms of inducing phagocytosis is CR3, which is also known as $\alpha_M\beta 2$, CD11/CD18 or Mac-1 (Uribe-Querol & Rosales, 2020).

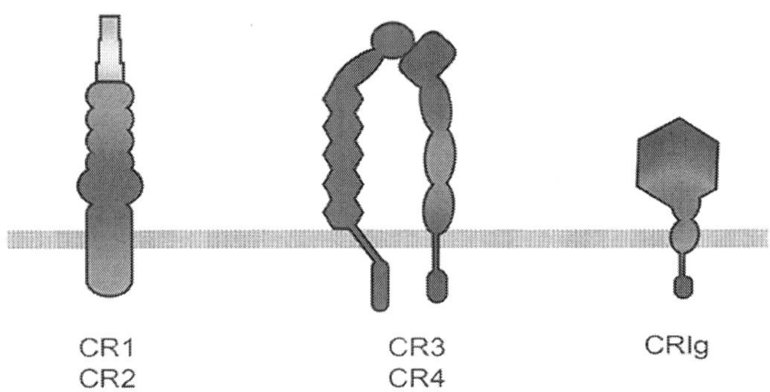

Figure 2.38. The three groups of complement receptors are CR1 and CR2, CR3 and CR4, and CRIg. (Source: Eileen Uribe-Querol & Carlos Rosales, Creative Commons License)

Efficient recognition of the target particle requires that numerous receptors on the membrane of the phagocyte must bind with several Ig molecules on the opsonized particle (see Figure 2.39). For this to happen, the receptors must move along the membrane in order for them to aggregate or crosslink and get activated. However, due to the short size of phagocytic receptors, their lateral diffusion and access to the ligand are obstructed by bigger transmembrane glycoproteins, like mucins, hyaluronan, or phosphatases. Many of these large glycoproteins are also bound to the underlying cytoskeleton so that they act like immobile fences and thus, interfere with the movement and clustering of phagocytic receptors (Uribe-Querol & Rosales, 2020).

FcγRscooperate with other receptors, such as activated integrins, to remove the larger glycoproteins from the membrane area in contact with the target particle. This enables FcγRs to diffuse freely along the membrane and engage more IgG molecules. CR3 can increase their affinity to their ligand after they receive a cytokine-mediated inside-out signal (i.e., a signal coming from inside the cell) from other receptors, which include FcγR, TLRs, or CD44, resulting in its cleavage into C3a and C3b (see Figure 2.39). Inside-out signaling involves the phosphorylation of tyrosine residues in the ITAM of clustered FcγRs by Src family kinases (SFKs). Then, spleen tyrosine kinase (Syk) docks to the phosphorylated ITAM and is also phosphorylated, either by SFKor by autophosphorylation. The activated Syk can recruit and activate molecules that remodel the cytoskeleton so that FcγRs can readily diffuse and bind the ligands forming a cluster in the process. Second messengers from the FcγR clusters initiate integrin activation via the small GTPase Rap1. These activated integrins, such as C3b, assume an extended conformation enabling them to bind several ligands and create a diffusion barrier so that larger glycoproteins like the phosphatase CD45 are kept away from phagocytic receptors. This conformational rearrangement also exposes more binding sites for the receptors. The second messengers involved in inside-out signaling can diffuse laterally so that the area of integrin activation is extended. In turn, this expands the zone of glycoprotein exclusion. Hence, integrin engagement acts as a progressive wave migrating ahead of the engaged Fcγ receptors. This wave bridges distant ligands leading to the formation of FcγR microclusters (Ostrowski *et al.*, 2016; Horiguchi *et al.*, 2018).

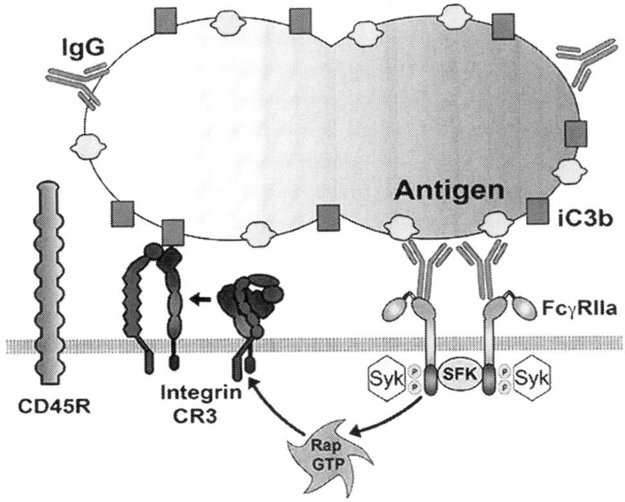

Figure 2.39. Most phagocytic receptors like FcγRIIa and integrin CR3 cooperate to bind the antigen to be engulfed. FcγRIIa receptors engage with sev-

eral IgG antibodies on the target so that these receptors aggregate triggering an inside-out signal that activates CR3 via the GTPase Rap. The activatedCR3 assumes an extended conformation allowing it to bind the target via the complement fragment C3b and to form a diffusion barrier that excludes larger transmembrane glycoproteins, such as CD45. (Source: Eileen Uribe-Querol & Carlos Rosales, Creative Commons License)

Phagocytes also extend their membranes to continuously probe their environment for foreign particles so that even scarcely opsonized or highly mobile targets are captured. This membrane protrusion requires actin polymerization mediated by Rho-family GTPases such as Rac and phosphoinositides. The mechanism by which this affects actin remodeling is discussed further in the next section (Flannagan *et al.*, 2010).

2.4.3.2 Internalization of the target particle

The binding of phagocytic receptors to their target initiates signaling pathways that result in remodeling of the actin cytoskeleton and lipids in the membrane so that the membrane in contact with the particle forms a concavity called the phagocytic cup. Then the two lips of the cup extend forming pseudopods that fuse around the particle to create an early phagosome inside the cell (Uribe-Querol & Rosales, 2020).

As described in the preceding section, FcγR clustering triggers SFK/Syk-mediated signaling pathway that initiates phagocytosis-related processes (see Figure 2.40). Activated Syk can phosphorylate several substrates involved in phagocytosis including the adaptor molecule linker for activation of T cells (LAT), phosphatidylinositol 3-kinase (PI 3-K), and phospholipase C gamma (PLCγ). Phosphorylation of LAT induces the docking of additional adapter molecules like Grb2 and Grb2-associated binder 2 (Gab2). Phosphorylated PI 3-K generates the lipid phosphatidylinositol-3,4,5-triphosphate (PIP_3) at the phagocytic cup. PIP_3regulates the activation of the GTPase Rac and contractile proteins such as myosin. The guanine nucleotide exchange factor (GEF) Vav also activates Rac and another GTPase Rho. Activated Rac plays a role in actin polymerization that drives pseudopod extension and in activation of other signaling molecules such as JNK and NF-κB. For its part, active PLCγ produces the second messengers inositol triphosphate (IP_3) and diacylglycerol (DAG). These induce calcium release from the ER and activation of protein kinase C (PKC), respectively. PKC, when activated, stimulates extracellular signal-regulated kinases ERK and p38 (Flannagan *et* al., 2010; Uribe-Querol & Rosales, 2020).

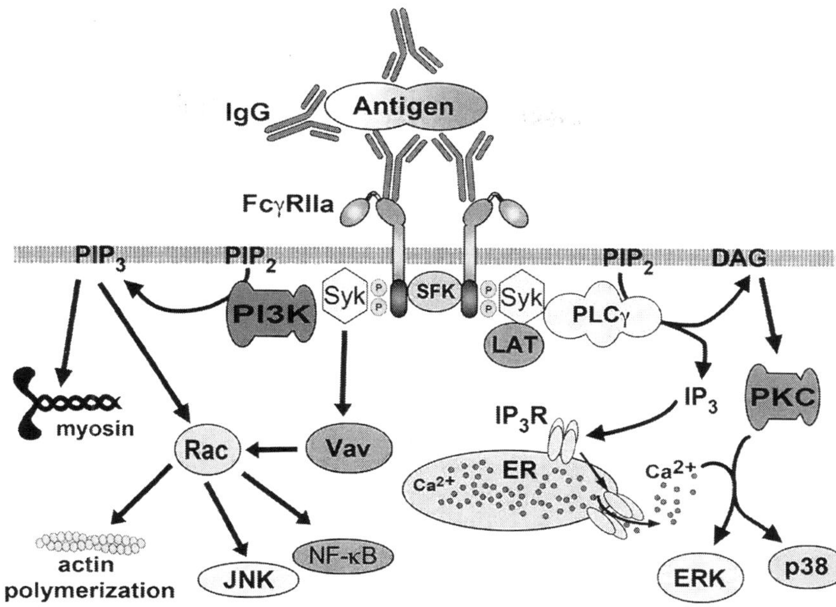

Figure 2.40. When FcγRIIa engages IgG antibodies on the target particle (antigen), it activates Src family kinase (SFK), which phosphorylates tyrosine residues in the ITAM (green domain) of the receptor. Then spleen tyrosine kinase (Syk) associates with the phosphorylated ITAMs and phosphorylates a signaling complex composed of the scaffold protein linker for activation of T cells (LAT) and other proteins, such as phospholipase C gamma (PLCγ). The latter produces inositoltriphosphate (IP_3) that induces calcium release from the endoplasmic reticulum (ER) and diacylglycerol (DAG) that activates protein kinase C (PKC). PKC activates extracellular signal-regulated kinases ERK and p38. Syk also recruits and activates phosphatidylinositol 3-kinase (PI_3K), which produces phosphatidylinositol-3,4,5-triphosphate (PIP_3) that regulates contractile proteins like myosin. PIP_3 and Vav activate the GTPase Rac, which is involved in the activation of the transcription factors, such as NF-κB and JNK. (Source: Eileen Uribe-Querol & Carlos Rosales, Creative Commons License)

CR3 initiates a type of phagocytosis that differs from the one mediated by FcγRs. Specifically, it is characterized by the sinking of the target particle into the cell membrane without the formation of pseudopods around the particle. Moreover, FcγR-initiated phagocytosis employs actin while CR3-initiated one involves actin and microtubules. Complement phagocytosis depends on the activated GTPase Rho to promote actin remodeling and polymerization via two mechanisms (see Figure 2.41). First, Rho activates

Rho kinase (ROCK), which phosphorylates and activates myosin II. This leads to the activation of the Arp2/3 complex, which promotes actin assembly at the phagocytic cup. Second, Rho induces the accumulation of mammalian diaphanous-related formin 1 (mDia1), which stimulates actin polymerization in the phagocytic cup. For its part, mDia1 binds directly to the microtubule-associated protein CLIP-170 and provides a link to the microtubule required for phagocytosis (Uribe-Querol & Rosales, 2020).

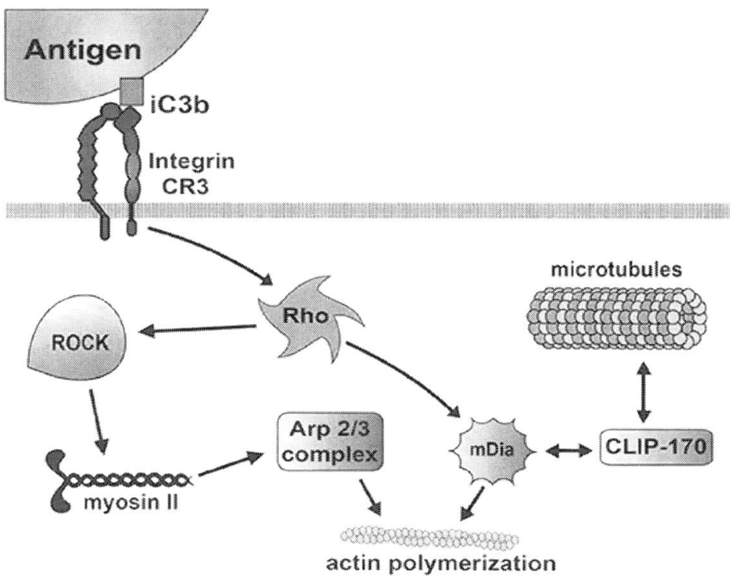

Figure 2.41. When the complement receptor 3 (CR3) binds the complement molecules iC3b on the target particle (antigen), a signaling pathway is initiated, which activates the GTPase Rho. Rho may activate Rho kinase (ROCK), which phosphorylates and activates myosin II leading to the accumulation of Arp2/3 complex that promotes actin polymerization at the phagocytic cup. Rho may also promote the accumulation of mammalian diaphanous-related formin 1 (mDia1) which drives actin polymerization and binds to CLIP-170 providing a link to microtubules. (Source: Eileen Uribe-Querol & Carlos Rosales, Creative Commons License)

2.4.3.3 Formation of phagosome

Once the pseudopods fuse around the ingested particle, the vesicle formed pinches off from the plasma membrane forming the phagosome. As the phagosome is being formed, its membrane changes in terms of lipid composition through the generation and degradation of various lipid molecules. In the case of Fcγ receptor-mediated phagocytosis, phosphatidylinositol-4,5-bisphosphate (PIP_2) initially accumulates at the phagocytic cup and then rapidly declines due to its phosphorylation by PI 3-K to produce PIP_3. This decrease in PIP_2level may facilitate actin disassembly, which is critical for the internalization of the target particle. Reduction of PIP_2 is also mediated by PLCγ, which produces DAG that induces PKCεfor enhanced phagocytosis (Uribe-Querol & Rosales, 2020).

Another change that occurs in the membrane during phagosome formation is actin remodeling that enables the membrane to protrude and cover the target. The first step for pseudopod formation is the disruption of the cortical cytoskeleton via the action of coronins, which are F-actin debranching proteins, and the F-actin severing proteins, cofilins and gelsolin. Coronin 1 debranches F-actin in the cortical region producing linear fibers that can be severed by cofilin and gelsolin. Secondly, pseudopods are formed by F-actin polymerization, which is induced by the actin-nucleating activity of the Arp2/3 complex and mDia1. The former produces branched actin filaments while the latter forms long, straight actin filaments at the phagocytic cup. Finally, actin at the base of the phagocytic cup disintegrates, which is thought to facilitate pseudopods to curve around the target, fuse, and then form the phagosome. This process involves both the termination of actin polymerization and activation of depolymerization, both of which seem to be regulated by PI 3-K, which produces PIP_3that deactivates GTPases. As a result, actin polymerization is inhibited (Uribe-Querol & Rosales, 2020).

The actin-binding proteins, myosins, seem to use their contractile activity to facilitate phagosome formation. In macrophages, class II and class IXb myosins are concentrated at the base of phagocytic cups, myosin Ic level increases at the site of phagocytic closure, and myosin V appears after phagosome closure. As the pseudopod extends, a tight ring of actin filaments moves from the bottom of the phagocytic cup to the top squeezing the particle to be engulfed. This contractile activity is dependent on myosin light-chain kinase (MLCK), which activates myosin II. The squeezing action pushes extra fluid out of the phagosomes. Myosin IX, similar to myosin II, is believed to be involved in the contractile activity of phagocytic cups. It may also function as a signaling molecule for actin reorganization since it contains a GTPase activation protein (GAP) domain that activates Rho. Myosin X is also recruited to the phagocytic cup in a PI 3-K-mediated process and is involved in pseudopod extension. Myosin Ic, which accumulates at the tip of the phagocytic cup, is implicated in generating the contraction force that closes the opening of the phagocytic cup in a purse string-like manner. Finally, myosin V, which appears in fully internalized phagosomes, is possibly responsible for the short-range movement of the newly formed phagosomes (Uribe-Querol & Rosales, 2020).

2.4.3.4 Maturation of phagosome

After the target particle is completely internalized and is now inside a phagosome, it undergoes a series of fusion and fission interactions with vesicles coming from the ER-Golgi complex (see Figure 2.42). This process is called phagosome maturation.

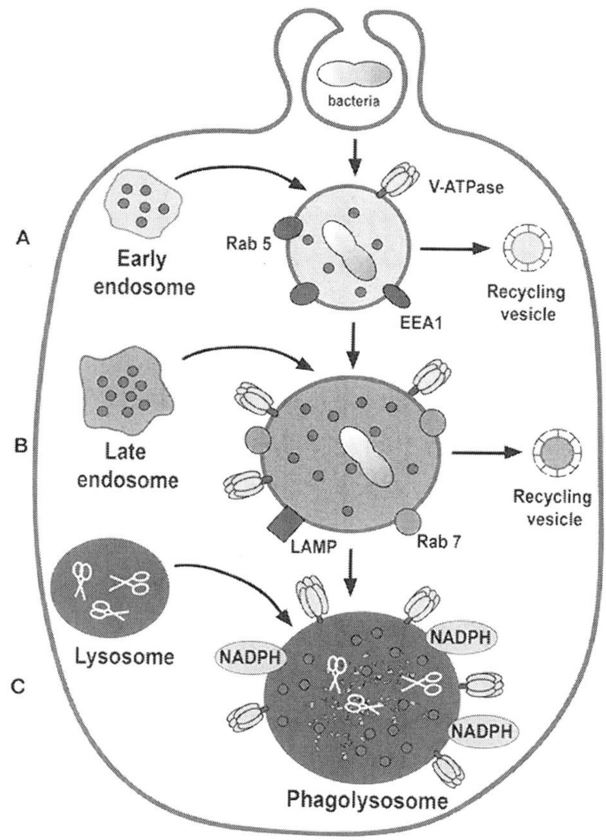

Figure 2.42. The phagosome undergoes a three-stage maturation process: early (A), late (B), and the phagolysosome formation (C). As the phagosome fuses with an endosome, its size remains the same because recycling vesicles also bud off from it. However, its membrane composition changes to include mol-

ecules that regulate membrane fusion, including Rab5, early endosome antigen 1 (EEA1), and Rab7. The interior of the phagosome also acidifies with the phagolysosome being the most acidic due to the action of the proton-pumping V-ATPase. The phagolysosome also acquires several degradative enzymes (scissors) and, the NADPH oxidase complex in its membrane initiates a pathway that produces the microbicide hypochlorous acid (Uribe-Querol & Rosales, 2020).

The new phagosome fuses with the early endosome to form an early phagosome (see Figure 2.42). This process is initiated by the small GTPase Rab5 that recruits early endosome antigen 1 (EEA1). EEA1 acts as a bridge between early endosomes and endocytic vesicles and mediates the recruitment of other proteins like Rab7. The phagosome maintains its size because as it fuses with several endosomes, recycling endosomes are simultaneously removed from the phagosome (Uribe-Querol & Rosales, 2020).

During the maturation process, Rab5 diminishes and Rab7 appears on the membrane. Rab7 promotes the fusion of the phagosome with the late endosome to form the late phagosome (see Figure 2.42). V-ATPase also accumulates on the phagosome membrane. This molecule causes the phagosome interior to acidify with a pH of 5.5-6.0 by translocating H^+ into the lumen of the phagosome. Lysosomal-associated membrane proteins (LAMPs) and luminal proteases, such as cathepsins and hydrolases, are also integrated into the membrane from the fusion with late endosomes (Uribe-Querol & Rosales, 2020).

Finally, the late phagosome combines with the lysosome to become phagolysosome, which can now degrade that ingested particle via several mechanisms (see Figure 2.42). First, many V-ATPase molecules accumulated on its membrane make its lumen very acidic with a pH as low as 4.5. Its membrane also has the NADPH oxidase complex, which produces ROS like superoxide (O_2^-)that is converted to H_2O_2. Myeloperoxidase catalyzes the reaction of H_2O_2. with Cl^- ions to form the potent antimicrobial substance hypochlorous acid. The phagolysosome also contains several hydrolases, such as cathepsins, proteases, lysozymes, and lipases, which can degrade the target particle (Uribe-Querol & Rosales, 2020).

Phagosome formation and maturation can be influenced by PAMPs like lipopolysaccharide and cytokines, including TNFα and IFNγ. During phagosome maturation, phagocytes are also triggered to produce cytokines to drive the migration and activation of other immune cells to the site of infection (Harris, 2021).

2.4.4 Inflammatory Response

Inflammation is initiated when innate immune cells detect an infection or tissue injury. It provides broad spectrum protection against such infection as well as establishes long-term adaptive immunity toward specific pathogens (Xiao, 2017). It is characterized by redness and warmth in the injured and infected site, which is caused by increased blood flow to the area, as well as swelling that is due to leakage of fluid, plasma proteins, and leukocytes from the blood vessels (Pober & Sessa, 2014) (see Figure 2.43). Chronic and uncontrolled inflammation due to overzealous immune responses causes severe tissue damage leading to major pathogenicity. Therefore, inflammation is tightly regulated by mechanisms that oversee its initiation, progression, and resolution (Xiao, 2017).

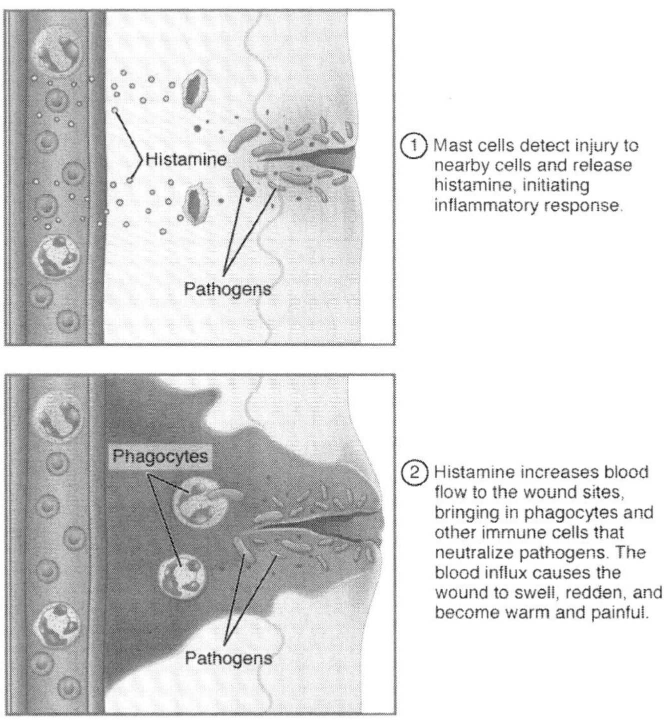

Figure 2.43. Inflammation is induced by the detection of foreign entities or tissue injury. Histamine is released causing increased blood flow that leads to the inflamed site becoming red and warm. The influx of immune cells leads to the characteristic swelling. (Source: Lindsay M. Biga *et al.*, Creative Commons License)

Histamine, prostaglandins, and nitric oxide cause vasodilation of blood vessels to increase blood flow and bring in circulating leukocytes (see Figure 2.43). In addition, histamine and leukotrienes act on endothelial cells to increase vascular permeability and enable plasma proteins and leukocytes to leave the circulation. Cytokines that include TNF and IL1 increase the levels of leukocyte adhesion molecules on endothelial cells to enhance the leakage of leukocytes from the blood vessels (Newton & Dixit, 2012).

2.4.4.1. Induction of pro-inflammatory mediators

The activation of various types of PRRs leading to the synthesis of various pro-inflammatory cytokines has been described in greater detail in section 2.3.1. This portion recapitulates the significance of receptor-ligand interaction in initiating the inflammatory response.

Intracellular or membrane PRRs on cells like macrophages, fibroblasts, mast cells, and DCs detect PAMPs or DAMPs. The binding of the ligands to the PRRs triggers a cascade of events that lead to the activation of transcription factors that control inflammatory responses. These transcription factors include NF-κB, IRFs, STAT, and AP1 families. The preferential stimulation of subsets of these transcription factors is influenced by the specific PRR activated by certain signals. As an example, activation of TLR3 and TLR4 is coupled with IRF3 activation resulting in the induction of IFN-β gene expression and the downstream IFN response. On the other hand, NF-κB and AP1 are stimulated in response to various inflammatory agonists ranging from microbial products to TNFα. However, these transcription factors are generally not activated by cytokines, like IFNγ, that activate STAT (Smale & Natoli, 2014). The activated PRRs, particularly TLRs, then oligomerize and assemble large complexes that induce the MyD88-mediated NF-κB or MAPK signaling pathways, which stimulate the release of cytokines that promote migration of leukocytes to the site of infection or injury. MAPK's JNK and p38α are activated downstream from TLR2 and TLR4. JNK regulates the activity of the transcription factor AP1 and drives the expression of pro-inflammatory cytokines like TNF. On the other hand, p38α mediates the activation of the transcription factors CREB and c/EBPβ and the expression of *Cxcl1* and *Cxcl2* chemokine genes, cytokine genes such as *IL10*, *IL12b*, *IL1a*, and *IL1b*, and genes for regulators of extracellular matrix remodeling (*Mmp13*) and cell adhesion (*Vcam1*). NF-κB transcription factors, meanwhile, induce the expression of a large number of pro-inflammatory genes (Newton & Dixit, 2012).

NF-κB also drives the expression of genes that regulate the duration and magnitude of the inflammatory response, which include *Tnfaip3* and *Nfkbia*. These genes are critical in ensuring that inflammation does not cause more tissue damage than the initial injury. *Tnfaip3* codes for the A20 deubiquitinating enzyme, which is thought to switch off TLR signaling and possibly functions as a tumor suppressor. *Nfkbia* encodes IκBα, which inhibits NF-κB signaling by forming a negative-feedback loop (Newton & Dixit, 2012).

TLR 7 and TLR9 activation triggers MyD88-mediated signaling cascade that induces the production not only of pro-inflammatory cytokines but also of IFNα and IFNβ in pDCs. These cytokines are essential to the anti-viral response. These type I IFNs (IFNI) are also induced by TRIF-dependent signaling following TLR3 and TLR4 activation (Newton & Dixit, 2012).

IFNI production is also triggered by the MAVS-mediated IRF activation as a result of the activation of RLRs like RIG-I and MDA5. Moreover, cytosolic DNA receptors, such as cGAS, AIM2, DDX41, Rad50, LRRFIP1, DNA-dependent activator of IRFs (DAI), as well as RNA sensors like IFN-induced protein with tetratricopeptide repeats 1 (IFIT1) are critical in stimulating antiviral immune response as mediated by the STING and MAVS pathways (Liu & Cao, 2016). NLRs like NOD1 and NOD2 also activate the NF-κB signaling cascade. Other types of NLRs likeNLRP1, NLRP3, and NLRC4 drive the production of pro-inflammatory cytokines, such as IL1β and IL18, and these are also triggered by certain PAMPs and DAMPs to nucleate signaling complexes called inflammasomes. The caspase 1 in the inflammasome complex, when activated, cleave its substrates, pro-IL1β and pro-IL18, leading to their subsequent maturation (Newton & Dixit, 2012).IL1β and IL18 are then secreted enabling a rapid pro-inflammatory response. Inflammasome activation also triggers pyroptosis, which serves to inhibit the intracellular replication of pathogens (de Zoete *et al*., 20140).

Another class of receptors involved in the inflammatory response is the FcR. FcRs on leukocytes recognize and engage complexes containing IgG or IgE antibodies. FcεRi is a high affinity receptor for IgE and it is expressed on mast cells. It has a key role in allergic reactions as its activation triggers intracellular granules to fuse with the plasma membrane so that the contents, i.e., preformed inflammatory mediators like histamine, serotonin, and proteases, are secreted extracellularly. Activated mast cells also release pro-inflammatory prostaglandins, leukotrienes, and cytokines (Newton & Dixit, 2012).

2.4.4.2 Amplification of the inflammatory response

IL1β and TNF promote NF-κB and MAPK activation amplifying the inflammation process. Binding of IL1β to IL1R activates MyD88-dependent signaling. On the other hand, TNF binds to TNF receptor 1 (TNF-RI) to enhance inflammation. TNF-RI is a transmembrane protein and TNF specifically binds with its cysteine-rich domains (CRD3) located extracellularly. Its cytoplasmic tail contains a death domain (DD) that recruits TRADD and the kinase RIP1. TRADD facilitates the binding of RIP1 to TNF-R1 and recruits TRAF2, which is an adaptor for the ubiquitin ligases cIAP1 and cIAP2. Recruitment of linear ubiquitin chain assembly complex (LUBAC) components, namely sharpin, HOIL-1, and HOIL-2, to the TNF-R1 complex is associated with cIAP activity. When RIP1 is ubiquitinated by cIAP1 and E2 UbcH5, it recruits NEMO and TAK1, eventually resulting in the activation of IKK. LUBAC mediates the linear ubiquitination of RIP1 and NEMO possibly contributing to the stability of the TNF-R1 signaling complex. In some cell types, ubiquitination of TRAF2 or cIAPs leads to IKK activation. IKKβ activation triggers NF-κB and Tp12-MEK1-ERK kinase cascade. The activation of Tp12 is linked to the activation of the MKK4-JNK pathway. Also, ERK activates the kinase MSK1 so that it phosphorylates RelA leading to the increased NF-κB-dependent transcription (Newton & Dixit, 2012).

AMPs like lipocalin-2 and activated complement components can also amplify the initial inflammatory response that was induced. As a result, a significantly increased transcription occurs in the immune cells so that high levels of inflammatory cytokines, chemokines, biogenic amines, and eicosanoids are produced within minutes to hours from the time PAMPs or DAMPs are sensed (Cronkite &Strutt, 2018).

Pro-inflammatory chemokines and activated complement recruit additional innate immune cells to the site of infection or injury. These cells surround the damaged or infected cells and release more pro-inflammatory cytokines. Neutrophils also release DNA nets to trap free extracellular pathogens, while NK cells lyse infected host cells. The innate inflammatory response and cellular swarm contain the pathogen until highly specific, activated cells of the adaptive immune system ultimately clear the infection (Cronkite & Strutt, 2018).

2.4.4.3 Increased cell adhesion

TNF and IL1β signaling, along with lipopolysaccharide detection, in endothelial cells can increase the expression of transmembrane proteins involved in cell adhesion, which include P-selectin, E-selectin, ICAM1 and VCAM1. Leukocyte glycoproteins like CD44, P-selectin glycoprotein ligand (PSGL1), and E-selectin ligand 1 (ESL1) bind the selectins to facilitate the rolling of leukocytes along the wall of blood vessels. Upon the attachment of either PSGL1 or CD44 to selectin, a signaling pathway is activated that causes the β2 integrins LFA1 and MAC1 on the leukocyte surface to shift to a more extended conformation. This increases their affinity for endothelial ICAM1 so that leukocyte rolling slows down. Chemokines and chemoattractants immobilized on the endothelial cell surface interact with G-protein-coupled receptors (GPCRs) to mediate further activation of β2 integrin leading to leukocyte arrest. Then, ICAM1 clustering on endothelial cells induces signals that facilitate leukocyte migration across the endothelium into the surrounding tissue (Newton & Dixit, 2012).

The complement protein C5aalso stimulates endothelial cells to upregulate the expression of cytokines, chemokines, and cell adhesion molecules like E-selectin, ICAM1, and VCAM1. It is produced by complement plasma proteases, which are activated by IgM and IgG-containing antibody complexes, pathogens coated with host mannose-binding lectin or C-reactive protein, or isolated pathogens. It can also be produced by non-complement proteases, such as thrombin and kallikrein that are components of the clotting system (Newton & Dixit, 2012).

2.4.4.4 Regulating the inflammatory response

Inflammation is strictly regulated to ensure that the vital functions of organs are not compromised by the immune response preventing pathological inflammation that may even lead to autoimmune diseases, inflammatory diseases, and cancer. Multi-level mechanisms are at play in dampening the inflammatory response.

The inflammatory response is initiated by PRR activation, which involves several thousand of genes. Therefore, the proper timing and setting of inflammation depend on the right combination of genes that are expressed and silenced. This is accomplished with the help of epigenetic mechanisms that regulate chromatin status and gene-specific transcription. Under healthy conditions, inactive enhancer sequences in the DNA are occupied by lineage-determining transcription factors known as pioneers, such as PU.1, and these sequences are marked with H3K4me1 (lysine 4 of histone 1 is monomethylated) and repressive H3K27me3 (lysine 27 of histone 3 is trimethylated). When TLR is activated, PU.1 allows the binding of signal-dependent transcription factors like NF-κB, IRFs, AP1, and STAT. It also promotes the removal of the H3K27me3 mark and the deposition of H3K27ac (lysine 27 of histone 3 is acetylated) resulting in the relaxation of chromatin structure. Enzymes that are involved in epigenetic mechanisms are also observed to influence inflammatory gene expression. In particular, the DNA methyltransferase Dnmt3a triggers histone deacetylase 9 (HDAC9) to deacetylase the kinase TBK1 leading to its activation that contributes to increased IFN production (Liu & Cao, 2016).

Unnecessary inflammation is also prevented by soluble as well as cell surface ligands. For instance, tissue-derived factors like surfactant proteins and mucins can inhibit the activation of DAMP receptors. DNA may also act as a ligand that activates the inhibitory DAMP receptor and the inhibitory cytosolic receptor. Another example is CD200, which ligates CD200Rs on the surface of monocytes and DCs inducing inhibitory signals. In lung APC, the release of mitochondrial H_2O_2suppresses the NF-κB signaling pathway. In addition, Tregs and tissue cells produce anti-inflammatory cytokines, such as IL10 and TGFβ, that downregulate inflammation (Cronkite & Strutt, 2018).

Moreover, certain nuclear receptors inhibit NF-κB activity or enhance NCoR/SMRT corepressor complex so that TLR4-induced NF-κB downstream genes are suppressed leading to limited inflammatory gene programs. These receptors include glucocorticoid receptor, peroxisome proliferator-activated receptor-gamma, liver X receptor, small heterodimer partner (SHP), and nuclear receptor subfamily 4 group A members (i.e., NR4A1 and NR4A2) (Liu & Cao, 2016).

The post-translational modification of PRRs can also regulate innate inflammatory signaling by affecting their function and activity. For example, removal of lysine 63-linked polyubiquitin chain of TRAF6 by A20 and TANKregulates TRAF6 activity resulting in inhibition of TLR signaling activation. On the other hand, phosphorylation of TRAF6 by germinal center kinase MST4 prevents oligomerization and autoubiquitination of TRAF6 so that inflammatory responses are inhibited (Liu & Cao, 2016).

Cyclic adenosine monophosphate, which is part of the ubiquitination of the dopamine (DA)/dopamine D1 receptor (DRD1) signaling pathway, binds to the inflammasome NLRP3 and recruits the E3 ubiquitin ligase MARCH7 leading to the ubiquitination and eventual degradation of NLRP3. This results in the downregulation ofNLRP3-dependent systemic inflammation (Liu & Cao, 2016).

Post-translational modification of signaling molecules, such as NEMO and IRF3, can also modulate type I IFN-mediated antiviral immunity. The tumor suppressor PTEN mediates the phosphorylation of serine 97 of IRF3, which controls IRF3 import into the nucleus. This contributes to the inhibition of IRF3 activation and IFNI production. Moreover, E3 ligase TRIM29 directly binds NEMO and triggers its ubiquitination and proteolytic degradation. As a result, NF-κB-mediated IRF3 signaling is also inhibited (Liu & Cao, 2016).

2.4.4.5 Imprinting of inflammatory response

Similar to the adaptive immune system, innate immunity can also be altered or trained by past infection, although it is not as specific as the memory immune responses characteristic of adaptive immunity. Most of the innate immune cells exhibit imprinting that leads to a generic and nonspecific heightened inflammatory response during secondary infection. Organs like the lungs remain in an "imprinted" state after the resolution of the infection or injury, and it may last for days, weeks, or even months. This can provide a degree of protection against unrelated pathogens (Cronkite & Strutt, 2018).

Innate immune cells can also be trained by the phagocytosis of apoptotic cells in the absence of infection. This is a normal process for cellular turnover that does not generate DAMPs. Nevertheless, it can imprint macrophages for heightened inflammatory responses providing nonspecific resistance to infection (Cronkite & Strutt, 2018).

The imprinted innate immunity is correlated with the presence of increased levels of activated macrophages, DCs, and other innate immune cells. In monocytes and macrophages, this state is maintained by long-term translational and epigenetic changes (Cronkite & Strutt, 2018).

CHAPTER REFERENCES

1. Abdallah, F., Mijouin, L. & Pichon, C. (2017). Skin Immune Landscape: Inside and Outside the Organism. *Mediators of Inflammation* 5095293. https://doi.org/10.1155/2017/5095293 (Accessed on 30 June 2021).
2. Afshar, M. & Gallo, R.L. (2013). Innate immune defense system of the skin. *Advances in Veterinary Dermatology*. https://doi: 10.1111/j.1365-3164.2012.01082.x
3. (Accessed on 6 July 2021).
4. Aktor, J. (2012). *Elsevier's Integrated Review Immunology and Microbiology (2nd edn.)*. Philadelphia: Elsevier Saunders. pp.43-51. https://doi.org/10.1016/B978-0-323-07447-6.00006-5 (Accessed on 11 May 2021).
5. Aktor, J.K. (2019). *Introductory Immunology: Basic Concepts for Interdisciplinary Applications (2nd edn.)*. London: Academic Press. pp.1-16. https://doi.org/10.1016/B978-0-12-816572-0.00001-2 (Accessed on 17 May 2021).
6. Alberts, B., Johnson, A., Lewis, J., Raff, M., Roberts, K. & Walter, P. (2002). *Molecular Biology of the Cell (4th edn.)*. New York: Garland Science.
7. Amarante-Mendes, G.P., Adjemian, S., Migliari Branco, L., Zanetti, L.C., Weinlich, R. & Bortoluci, K.R. (2018). Pattern Recognition Receptors and the Host Cell Death Molecular Machinery. *Frontiers in Immunology* 9:2379. https://doi: 10.3389/fimmu.2018.02379 (Accessed on 1 June 2021).
8. American Cancer Society (2019). Cytokines and Their Side Effects. https://www.cancer.org/treatment/treatments-and-side-effects/treatment-types/immunotherapy/cytokines.html (Accessed on 18 May 2021).
9. Amin, K. (2012). The role of mast cells in allergic inflammation. *Respiratory Medicine* 106(1): 9-14. https://doi.org/10.1016/j.rmed.2011.09.007 (Accessed on 18 May 2021).
10. Ardain, A., Marakalala, M.J. & Leslie, A. (2019). Tissue-resident innate immunity in the lung. *Immunology* 159 (3): 245-256. https://doi.org/10.1111/imm.13143 (Accessed on 29 July 2021).
11. Bals, R. & Hiemstra, P.S. (2004). Innate immunity in the lung: how epithelial cells fight against respiratory pathogens. *European Respiratory Journal* 23: 327-333. https://doi:10.1183/09031936.03.00098803 (Accessed on 28 July 2021).

12. Bezuidenhout, J. & Schneider, J.W. (2009). Pathology and pathogenesis of tuberculosis. In: *Tuberculosis: A Comprehensive Clinical Reference*; Schaaf, H.S., Zumla, A.I., Grange, J.M., Raviglione, M.C., Yew, W.W., Starke, J.R., Pai, M. & Donald, P.R (Ed.). United States: Elsevier, Inc. pp. 117-128. https://doi.org/10.1016/B978-1-4160-3988-4.00012-3 (Accessed on 7 May, 2021).
13. Borden, E.C., Sem, G.C., Ue, G., Silverman, R.H., Ransohoff, R.M., Foster, G.R. & Stark, G.R. (2007). Interferons at age 50: past, current and future impact on biomedicine. *Nature Reviews Drug Discovery* 6: 975-990. https://doi.org/10.1038/nrd2422 (Accessed on 31 May 2021).
14. Borden, E.C., Silverman, R.H. & Sen, G. (2008). Interferons. In: *the Molecular Basis of Cancer (3rd edn.)*; Mendelsohn, J., Howley, P.M., Israel, M.A., Gray, J.W. & Thompson, C.B. Philadelphia: Elsevier Inc. pp. 621-633. https://doi.org/10.1016/B978-141603703-3.10052-4 (Accessed on 31 May 2021).
15. Brody, L. C. (2021). *Lymphocyte*. https://www.genome.gov/genetics-glossary/Lymphocyte (Accessed on 5 May 2021).
16. Buckland, K. (2021). Eosinophils. *British Society for Immunology*. https://www.immunology.org/public-information/bitesized-immunology/cells/eosinophils (Accessed on 4 May 2021).
17. Castell-Rodríguez, A., Piñón-Zárate, G., Herrera-Enríquez, M., Jarquín-Yáñez, K. & Medina-Solares, I. (2017). Dendritic Cells: Location, Function, and Clinical Implications. In: *Biology of Myelomonocytic Cells*; Ghosh, A. (Ed.). IntechOpen. https://doi: 10.5772/intechopen.68352 (Accessed on 14 May 2021).
18. Chiffoleau, E. (2018). C-Type Lectin-Like Receptors As Emerging Orchestrators of Sterile Inflammation Represent Potential Therapeutic Targets. *Frontiers in Immunology* 9:227. https://doi: 10.3389/fimmu.2018.00227 (Accessed on 22 June 2021).
19. Chirumbolo, S., Bjørklund, G., Sboarina, A. & Vella, A. (2018). The role of basophils as innate immune regulatory cells in allergy and immunotherapy. *Human Vaccines & Immunotherapeutics* 14(4): 815-831. https://doi: 10.1080/21645515.2017.1417711 (Accessed on 27 April 2021).
20. Chiu, S. & Bharat, A. (2016). Role of monocytes and macrophages in regulating immune response following lung transplantation. *Current Opinion in Organ Transplantation* 21(3): 239-245. https:// doi: 10.1097/MOT.0000000000000313 (Accessed on 10 May 2021).

21. Coates, M., Blanchard, S. & MacLeod, A.S. (2018). Innate antimicrobial immunity in the skin: a protective barrier against bacteria, viruses, and fungi. *PLoS Pathogens* 14(12): e1007353. https://doi.org/10.1371/journal.ppat.1007353 (Accessed on 24 June 2021).
22. Cronkite, D.A. & Strutt, T.M. (2018). The Regulation of Inflammation by Innate and Adaptive Lymphocytes. *Journal of Immunology Research* 1467538.https://doi.org/10.1155/2018/1467538 (Accessed on 1 September 2021).
23. Dafny, N., Yang, P.B. & Brod, S.A. (2004). Interferons. In: *Encyclopedia of Endocrine Diseases*; Martini, L. (Ed.). Philadelphia: Elsevier, Inc. pp. 53-59. https://doi.org/10.1016/B0-12-475570-4/00764-2 (Accessed on 31 May 2021).
24. De Kleer, I., Willems, F., Lambrecht, B. & Goriely, S. (2014). Ontogeny of myeloid cells. *Frontiers in Immunology* 5:423. https://doi: 10.3389/fimmu.2014.00423 (Accessed on 14 May 2021).
25. de Zoete, M.R., Palm, N.W., Zhu, S. & Flavell, R.A. (2014). Inflammasomes. *Cold Springs Harbor Perspectives in Biology*. https://doi: 10.1101/cshperspect.a016287 (Accessed on 8 September 2021).
26. Dixit, E. & Kagan, J.C. (2013). Intracellular Pathogen Detection by RIG-I-Like Receptors. In: *Advances in Immunology*Vol. 117; Alt, F.W. (Ed.). San Diego: Academic Press. pp. 99-125. https://doi.org/10.1016/B978-0-12-410524-9.00004-9 (Accessed on 10 June 2021).
27. Drouin, M., Saenz, J. & Chiffoleau, E. (2020). C-Type Lectin-Like Receptors: Head or Tail in Cell Death Immunity. *Frontiers in Immunology* 11:251. https://doi: 10.3389/fimmu.2020.00251 (Accessed on 22 June 2021).
28. Espinoza, V.E. & Emmady, P.D. (2020). Histology, Monocytes. In: *StatPearls [Internet]*. Florida: StatPearls Publishing.https://www.ncbi.nlm.nih.gov/books/NBK557618/ (Accessed on 10 May 2021).
29. Falcone, F.H., Haas, H. & Gibbs, B.F. (2000). The human basophil: a new appreciation of its role in immune responses. *Blood* 96(13): 4028-4038. https://doi.org/10.1182/blood.V96.13.4028 (Accessed on 30 April 2021).
30. Farooqui, A.A. (2018). *Ischemic and Traumatic Brain and Spinal cord Injuries: Mechanisms and Potential Therapies*. London: Academic Press. pp. 41-87. https://doi.org/10.1016/B978-0-12-813596-9.00002-X (Accessed on 18 May 2021).

31. Ferreira, V.L., Borba, H.H.L., Bonetti, A.d.F., Leonart, L.P. & Pontarolo, R. (2018). Cytokines and Interferons: Types and Functions. In: *Autoantibodies and Cytokines*; Khan, W.A. (Ed.). IntechOpen. https://doi:10.5772/intechopen.74550 (Accessed on 1 June 2021).
32. Flannagan, R.S.;, Harrison, R.E., Yip, Christopher M., Jaqaman, K. & Grinstein, S. (2010). Dynamic macrophage "probing" is required for the efficient capture of phagocytic targets. *Journal of Cell Biology* 191(6): 1205-1218. https://doi.org/10.1083/jcb.201007056 (Accessed on 2 September 2021).
33. Frey, A., Lunding, L.P., Ehlers, J.C., Weckmann, M., Zissler, U.M. & Wegmann, M. (2020). More Than Just a Barrier: The Immune Functions of the Airway Epithelium in Asthma Pathogenesis. *Frontiers in Immunology* 11: 761. https://doi: 10.3389/fimmu.2020.00761 (Accessed on 3 August 2021).
34. Gani, Z. (2021). Complement System. *British Society for Immunology*. https://www.immunology.org/public-information/bitesized-immunology/systems-and-processes/complement-system (Accessed on 20 September 2021).
35. Ganz, T. (2002). Epithelia: Not just physical barriers. *Proceedings of the National Academy of Sciences of the United States of America.* 99(6): 3357-3358. https://doi.org/10.1073/pnas.072073199 (Accessed on 26 April 2021).
36. Garrett, W.S. & Mellman, I. (2001). Studies of endocytosis. In: *Dendritic Cells (2nd edn.): Biology and Clinical Applications*; Lotze, M.T. & Thomson, A.W. (Ed.). London: Academic Press. pp. 213-229. https://doi.org/10.1016/B978-012455851-9/50055-9 (Accessed on 26 August 2021).
37. Geijenbeek, T.B.H. & Gringhuis, S.I. (2009). Signalling through C-type lectin receptors: shaping immune responses. *Nature Reviews Immunology* 9: 465-479. https://doi.org/10.1038/nri2569 (Accessed on 23 June 2021).
38. Germic, N., Frangez, Z., Yousefi, S. & Simon H.-U. (2109). Regulation of the innate immune system by autophagy: monocytes, macrophages, dendritic cells and antigen presentation. *Cell Death & Differentiation* 26: 715-727. https://doi.org/10.1038/s41418-019-0297-6 (Accessed on 11 May 2021).
39. Gibbs, B.F. (2008). Basophils as Key Regulators of Allergic Inflammation and Th2-type Immunity. *Wood Allergy Organization Journal* 1(7):

123-128. https://doi.org/10.1097/WOX.0b013e31817a76fb (Accessed on 3 May 2021).

40. Gravallese, E.M. & Monach, P.A. (2015). The rheumatoid joint: synovitis and tissue destruction. In: *Rheumatology (6th edn.)*; Hochberg, M.C., Silman, A.J., Smolen, J.S., Weinblatt, M.E. & Weisman, M.H. (Eds.). Philadelphia: Elsevier Moby. pp.768-784.https://doi.org/10.1016/B978-0-323-09138-1.00094-2 (Accessed on 24 May 2021).

41. Graves, D.T. & Jiang, Y. (1995). Chemokines, a family of chemotactic cytokines. *Critical Reviews in Oral Biology and Medicine* 6(2): 109-118. https://doi: 10.1177/10454411950060020101 (Accessed on 24 May 2021).

42. Guttman-Yassky, E., Zhou, L. & Krueger, J.G. (2019). The skin as an immune organ: Tolerance versus effector responses and application to food allergy and hypersensitivity reactions. *The Journal of Allergy and Clinical Immunology* 144(2): 362-374. https://doi.org/10.1016/j.jaci.2019.03.021 (Accessed on 5 July 2021).

43. Hartl, D., Tirouvanziam, R., Laval, J., Greene, C.M., Habiel, D., Shrama, L., Yildirim, A.Ö., Dela Cruz, C.S. & Hogaboam, C.M. (2018). Innate Immunity of the Lung: From Basic Mechanisms to Translational Medicine. *Journal of Innate Immunity* 10: 487-501. https://doi.org/10.1159/000487057 (Accessed on 29 July 2021).

44. Harris, J. (2021). Phagocytosis. *British Society for Immunology*. https://www.immunology.org/public-information/bitesized-immunology/systems-and-processes/phagocytosis (Accessed on 26 August 2021).

45. Hiemstra, P.S., Mc/cray, P.B., Jr. & Bals, R. (2015). The innate immune function of airway epithelial cells in inflammatory lung disease. *European Respiratory Journal* 45: 1150-1162. https://doi:10.1183/09031936.00141514 (Accessed on 27 July 2021).

46. Horiguchi, H., Loftus, T.J., Hawkins, R.B., Raymond, S.L., Stortz, J.A., Hollen, M.K., Weiss, B.P., Miller, E.S., Bihorac, A., Larson, S.D., Mohr, A.M., Brakenridge, S.C., Tsujimoto, H., Ueno, H., Moore, F.A., Moldawer, L.L., Efron, P.A. & The Sepsis and Critical Illness Research Center Investigators (2018). Innate Immunity in the Persistent Inflammation, Immunosuppression, and Catabolism Syndrome and Its Implications for Therapy. *Frontiers in Immunology* 9: 595. https://doi:10.3389/fimmu.2018.00595 (Accessed on 1 September 2021).

47. Hughes, C.E. & Nibbs, R.J.B. (2018). A guide to chemokines and their receptors. *The FEBS Journal* 285(16): 2944-2971. https://doi: 10.1111/febs.14466 (Accessed on 24 May 2021).
48. Hutcheon, C., Paulvannan, P. & Subramanian, N. (2016). Cytoplasmic Sensing in Innate Immunity. In *Encyclopediaof Cell Biology Vol. 13*. pp. 710-726. https://doi.org/10.1016/B978-0-12-394447-4.30108-0 (Accessed on 21 June 2021).
49. Janeway, C.A., Jr., Travers, P., Walport, M. & Shlomchik, M.J. (2001). Immunobiology: The Immune System in Health and Disease (5th edn.). New York: Garland Science.https://www.ncbi.nlm.nih.gov/books/NBK27156/ (Accessed on 17 May 2021).
50. Kato, H. & Fujita, T. (2016). Cytoplasmic Viral RNA Sensors: RIG-I-Like Receptors. In: *Encyclopedia of Immunobiology* Vol.2. pp. 352-359.https://doi.org/10.1016/B978-0-12-374279-7.02005-1 (Accessed on 10 June 2021).
51. Kato, J. & Svennson, C.I. (2015). Role of Extracellular Damage-Associated Molecular Pattern Molecules (DAMPs) as Mediators of Persistent Pain. In: *Progress in Molecular Biology and Translational Science Vol. 131: Molecular and Cell Biology of Pain*; Price, T.J. & Dussor, G. (Eds.). Massachusetts: Academic Press. pp. 251-279. https://doi.org/10.1016/bs.pmbts.2014.11.014 (Accessed on 2 June 2021).
52. Kawai, T. & Akira, S. (2009). The roles of TLRs, RLRs and NLRs in pathogen recognition. *International Immunology* 21(4): 317-337. https://doi: 10.1093/intimm/dxp017 (Accessed on 9 June 2021).
53. Kawasaki, T. & Kawai, T. (2014). Toll-like receptor signaling pathways. *Frontiers in Immunology* 5:461. https://doi: 10.3389/fimmu.2014.00461 (Accessed on 7 June 2021).
54. Kerrigan, A.M. & Brown, G.D. (2009). C-type lectins and phagocytosis. *Immunobiology* 214(7): 562-575. https://doi.org/10.1016/j.imbio.2008.11.003 (Accessed on 22 June 2021).
55. Kim, H.J., Kim, C.-H., Kim, M.-J., Ryu, J.-H., Seong, S.Y., Kim, S., Lim, S.J., Holtzman, M.J. & Yoon, J.H. (2015). The Induction of Pattern-Recognition Receptor Expression against Influenza A Virus through Duox2-Derived Reactive Oxygen Species in Nasal Mucosa. *American Journal of Respiratory Cell and Molecular Biology* 53(4): 525-535. https://doi: 10.1165/rcmb.2014-0334OC (Accessed on 29 July 2021).

56. Kobayashi, S.D., Malachowa, N. & DeLeo, F.R. (2018). Neutrophils and Bacterial Immune Evasion. *Journal of Innate Immunity* 10:432-441. https://doi.org/10.1159/000487756 (Accessed on 27 April 2021).
57. Kobayashi, T., Naik, S. & Nagao, K. (2019). Choreographing Immunity in the Skin Epithelial Barrier. *Immunity* 50(3): 552-565. https://doi.org/10.1016/j.immuni.2019.02.023 (Accessed on 27 April 2021).
58. Koyasu, S. & Moro, K. (2012). Role of innate lymphocytes in infection and inflammation. *Frontiers in Immunology* 3:101. https://doi: 10.3389/fimmu.2012.00101 (Accessed on 5 May 2021).
59. Kumar, V. (2020). Pulmonary Innate Immune Response Determines the Outcome of Inflammation During Pneumonia and Sepsis-Associated Acute Lung Injury. *Frontiers in Immunology* 11: 1722. https://doi: 10.3389/fimmu.2020.01722 (Accessed on 2 August 2021).
60. Kumar, V. & Sharma, A. (2010). Neutrophils: Cinderella of innate immune system. *International Immunopharmacology* 10(11): 1325-1334. https://doi.org/10.1016/j.intimp.2010.08.012 (Accessed on 28 April 2021).
61. Krystel-Whittermore, M., Dileepan, K.N. & Wood, J.G. (2016). Mast Cell: a Multi-Functional Master Cell. *Frontiers in Immunology* 6:620. https://doi: 10.3389/fimmu.2015.00620 (Accessed on 18 May 2021).
62. Larsen, S.B., Cowley, C.J. & Fuchs, E. (2020). Epithelial cells: liaisons of immunity. *Current Opinion in Immunology* 62: 45-53. https://doi.org/10.1016/j.coi.2019.11.004 (Accessed on 12 July 2021).
63. Leiva-Juárez, M.M., Kolls, J.K. & Evans, S.E. (2018). Lung epithelial cells: therapeutically inducible effectors of antimicrobial defense. *Mucosal Immunology* 11: 21-34. https://doi.org/10.1038/mi.2017.71 (Accessed on 2 August 2021).
64. Lewis, D.E. & Blutt, S.E. (2019). Organization of the Immune System. In: *Clinical Immunology: Principles and Practice (5th edn.)*; Rich, R.R., Fleisher, T.A., Shearer, W.T., Schroeder, H.W., Jr., Frew, A.J. & Weyand, C.M. (Ed.). United Kingdom: Elsevier Ltd. pp.19-38.e1. https://doi.org/10.1016/B978-0-7020-6896-6.00002-8 (Accessed on 5 May 2021).
65. Lin, C.D., Zhang, Y.H., Zhang, K., Zheng, Y.J., Lu, L., Chang, H.S., Yang, H., Yang, Y.R., Wan, Y.Y., Wang, S.H., Yuan, M.Y., Yan, Z.J., Zhang, R.G., He, Y.N., Ge, G.X., Wu, D. & Chen, J.F. (2019). Fever Promotes T lymphocyte Trafficking via a Thermal Sensory Pathway

Involving Heat Shock Protein 90 and α4 Integrins. *Immunity* 50(1): P137-151. https://doi.org/10.1016/j.immuni.2018.11.013 (Accessed on 9 August 2021).

66. Liu, J. & Cao, X. (2016). Cellular and molecular recognition of innate inflammatory responses. *Cellular & Molecular Immunology* 13: 711-721. https://doi.org/10.1038/cmi.2016.58 (Accessed on 2 September 2021).
67. Liu, Q., Yu, Z., Tian, F., Zhao, J., Zhang, H., Zhai, Q. & Chen, W. (2020). Surface components and metabolites of probiotics for regulation of intestinal epithelial barrier. *Microbial Cell Factories* 19:23. https://doi.org/10.1186/s12934-020-1289-4 (Accessed on 13 July 2021).
68. Marshall, J.S., Warrington, R., Watson, W. & Kim. H.L. (2018). An introduction to immunology and immunopathology. *Allergy, Asthma & Clinical Immunology* 14:49. https://doi.org/10.1186/s13223-018-0278-1 (Accessed on 26 April 2021).
69. Matejuk, A. (2018). Skin Immunity. *Archivum Immunologiae Et Therapiae Experimentalis* 66(1): 45-54. https://doi: 10.1007/s00005-017-0477-3 (Accessed on 25 June 2021).
70. McGhee, J.R. & Fujihashi, K. (2012). Inside the Mucosal Immune System. *PLoS Biology* 10(9): e1001397. https://doi.org/10.1371/journal.pbio.1001397 (Accessed on 19 July, 2021).
71. Merad, M., Sathe, P., Helft, J., Miller, J. & Mortha, A. (2013). The Dendritic Cell Lineage: Ontogeny and Function of Dendritic Cells and Their Subsets in the Steady State and the Inflamed Setting. *Annual Review of Immunology* 31: 563-604. https://doi.org/10.1146/annurev-immunol-020711-074950 (Accessed on 11 May 2021).
72. Mnich. M.E., van Dalen, R. & van Sorge, N.M. (2020). C-Type Lectin Receptors in Host Defense Against Bacterial Pathogens. *Frontiers in Cellular and Infection Microbiology* 10:309. https://doi: 10.3389/fcimb.2020.00309 (Accessed on 18 June 2021).
73. Moens, E. & Veldhoen, M. (2002). Epithelial barrier biology: good fences make good neighbours. *Immunology* 135(1):1-8. https://doi: 10.1111/j.1365-2567.2011.03506.x (Accessed on 27 April 2021).
74. Mogensen, T.H. (2009). Pathogen Recognition and Inflammatory Signaling in Innate Immune Defenses. *Clinical Microbiology Reviews* 22(2): 240-273. https://doi: 10.1128/CMR.00046-08 (Accessed on 2 June 2021).

75. Muñoz-Carrillo, J.L, Castro-García, F.P., Gutiérrez-Coronado, O., Moreno-García, M.A. & Contreras-Cordero, J.F. (2017). Physiology and Pathology of Innate Immune Response Against Pathogens. In: *Physiology and Pathology of Immunology*; Rezai, N. (Ed.). IntechOpen. https://www.intechopen.com/books/physiology-and-pathology-of-immunology/physiology-and-pathology-of-innate-immune-response-against-pathogens (Accessed on 7 June 2021).
76. Muñoz-Carrillo, J.L, Contreras-Cordero, J.F., Gutiérrez-Coronado, O., Villalobos-Gutiérrez, P.T., Ramos-Garcia, L.G. & Hernández-Reyes, V.E. (2018). Cytokine Profiling Plays a Crucial Role un Activating Immune System to Clear Infectious Pathogens. In: *Immune response Activation and Immunomodulation* Tyagi, R.K. & Bisen, P.S. IntechOpen. https://doi:10.5772/intechopen.80843 (Accessed on 20 May 2021).
77. Murphy, P.M. (2008). Chemokines and chemokine receptors. In: *Clinical Immunology: Principles and Practice (3rd edn.)*; Rich, R.R., Fleisher, T.A., Shearer, W.T., Schroeder, H.W., Jr., Frew, A.J. & Weyand, C.M. (Eds.). pp. 173-196. Philadelphia: Elsevier Mosby. https://doi.org/10.1016/B978-0-323-04404-2.10011-9 (Accessed on 24 May 2021).
78. National Cancer Institute, 2021. *Mast cell*. https://www.cancer.gov/publications/dictionaries/cancer-terms/def/mast-cell (Accessed on 17 May 2021).
79. Newton, K. & Dixit, V.M. (2012). Signaling in Innate Immunity and Inflammation. *Cold Spring Harbor Perspectives in Biology* 4(3): a006049. https://doi: 10.1101/cshperspect.a006049 (Accessed on 30 August 2021).
80. Okumura, R. & Takeda, K. (2017). Role of intestinal epithelial cells in the maintenance of gut homoeostasis. *Experimental & Molecular Medicine* 49: e338. https://doi.org/10.1038/emm.2017.20 (Accessed on 15 July 2021).
81. Ostrowski, P.P., Grinstein, S. & Freeman, S.A. (2016). Diffusion Barriers, Mechanical Forces, and the Biophysics of Phagocytosis. *Developmental Call* 38(2): 135-146. https://doi.org/10.1016/j.devcel.2016.06.023 (Accessed on 19 August 2021).
82. Parigi, S.M., Eldh, M., Larssen, P., Gabrielsson, S. & Villalanca, E.J. (2015). Breast milk and solid food shaping intestinal immunity. *Frontiers*

in Immunology 6: 415. https://doi: 10.3389/fimmu.2015.00415 (Accessed on 13 July 2021).

83. Parker, D. & Prince, A. (2011). Innate Immunity in the Respiratory Epithelium. *American Journal of Respiratory Cell and Molecular Biology* 45(2): 189-201. https://doi: 10.1165/rcmb.2011-0011RT (Accessed on 27 July 2021).
84. Pober, J.S. & Sessa, W.C. (2014). Inflammation and the Blood Microvascular System. *Cold Spring Harbor Perspectives in Biology*. https://doi: 10.1101/cshperspect.a016345(Accessed on 8 September 2021).
85. Prince, L., Parker, L. & Sabroe, I. (2020). Pathogen Recognition Receptors and the Lung. In: *Reference Module in Biomedical Sciences*. https://doi.org/10.1016/B978-0-08-102723-3.00008-1 (Accessed on 11 June 2021).
86. Proudfoot, A.E.I., Johnson, Z., Bonvin, P. & Handel, T.M. (2017). Glycosaminoglycan Interactions with Chemokines Add Complexity to a Complex System. *Pharmaceuticals* 10(3): 70. https://doi.org/10.3390/ph10030070 (Accessed on 31 May 2021).
87. Quaresma, J.A.S. (2019). Organization of the Skin Immune System and Compartmentalized Immune Responses in Infectious Diseases. *Clinical Microbiology Reviews* 32(4): e00034-18.https://doi.org/10.1128/CMR.00034-18 (Accessed on 5 July 2021).
88. Rehwinkel, J. & Gack, M.U. (2020). RIG-I-like receptors: their regulation and roles in RNA sensing. *Nature Reviews Immunology* 20: 537-551. https://doi.org/10.1038/s41577-020-0288-3 (Accessed on 9 June 2021).
89. Reikine, S., Nguyen, J.B. & Modis, Y. (2014). Pattern recognition and signaling mechanisms of RIG-I and MDA5. *Frontiers in Immunology* 5:342. https://doi: 10.3389/fimmu.2014.00342 (Accessed on 16 June 2021).
90. Rhodes, J.W., Tong, O., Harman, A.N. & Turville, S.G. (2019). Human Dendritic Cell Subsets, Ontogeny, and Impact on HIV Infection. *Frontiers in Immunology* 10:1088. https://doi: 10.3389/fimmu.2019.01088 (Accessed on 13 May 2021).
91. Salomon, B.L., Leclerc, M., Tosello, J., Ronin, E., Piaggio, E. & Cohen, J.L. (2018). Tumor Necrosis Factor α and Regulatory T Cells in Oncoimmunology. *Frontiers in Immunology* 9:444. https://doi: 10.3389/fimmu.2018.00444 (Accessed on 1 June 2021).

92. Seadler, B.D., Toro, F. & Sharma, S. (2021). Physiology, Alveolar Tension. *StatPearls*. https://www.ncbi.nlm.nih.gov/books/NBK539825/ (Accessed on 22 July 2021).

93. Shamri, R., Xenakis, J.J. & Spencer, L.A. (2011). Eosinophils in innate immunity: an evolving story. *Cell and Tissue Research* 343: 57-83. https://doi: 10.1007/s00441-010-1049-6 (Accessed on 3 May 2021).

94. Sharma, L., Feng, J., Britto, C.J. & Dela Cruz, C.S. (2020). Mechanisms of Epithelial Immunity Evasion by Respiratory Bacterial Pathogens. *Frontiers in Immunology* 11: 91. https://doi: 10.3389/fimmu.2020.00091 (Accessed on 28 July 2021).

95. Sheshachalam, A., Srivastava, N., Mitchell, T., Lacy, P. & Eitzen, G. (2014). Granule protein processing and regulated secretion in neutrophils. *Frontiers in Immunology* 5:448. https://doi.org/10.3389/fimmu.2014.00448 (Accessed on 27 April 2021).

96. Silva, F.A.R., Rodrigues, B.L., Ayrizono, M.d.L.S. & Leal, R.F. (2016). The Immunological Basis of Inflammatory Bowel Disease. *Gastroenterology Research and Practice* 2097274. https://doi.org/10.1155/2016/2097274 (Accessed on 14 July 2021).

97. Smale, S.T. & Natoli, G. (2014). Transcriptional Control of Inflammatory Responses. *Cold Spring Harbor Perspectives in Biology*. https://doi: 10.1101/cshperspect.a016261 (Accessed on 8 September 2021).

98. Sue, P.K., Meir, M. & de la Morena, M. (2018). Immunologic Development and Susceptibility to Infection. In: *Principles and Practice of Pediatric Infectious Diseases (5th edn.)*; Long, S.S., Prober, C.G. & Fischer, M. (Eds.). Philadelphia: Elsevier, Inc. pp.85-93.e3. https://doi.org/10.1016/B978-0-323-40181-4.00009-8 (Accessed on 1 June 2021).

99. Takeda, K. & Akira, S. (2005). Toll-like receptors in innate immunity. *International Immunology* 17(1): 1-14. https://doi.org/10.1093/intimm/dxh186 (Accessed on 9 June 2021).

100. Torres, J., Enriquez-de-Salamanca, A., Fernández, I., Rodríguez-Ares, M.T., Quadrado, M.J., Murta, J., Benítez del Castillo, M.T., Stern, M.E. & Calonge, M. (2011). Activation of MAPK signaling Pathway and NF-κB Activation in Pterygium and Ipsilateral Pterygium-Free Conjunctival Specimens. *Investigative Ophthalmology & Visual Science* 52(5842-5852). https://doi.org/10.1167/iovs.10-6673 (Accessed on 8 June 2021).

101. Travers, J. & Rothenberg, M.E. (2015). Eosinophils in mucosal immune responses. *Mucosal Immunology* 8:464-475. https://doi.org/10.1038/mi.2015.2 (Accessed on 4 May 2021).

102. Uribe-Querol, E. & Rosales, C. (2020). Phagocytosis: Our Current Understanding of a Universal Biological Process. *Frontiers in Immunology* 11:1066. https://doi: 10.3389/fimmu.2020.01066 (Accessed on 10 August 2021).

103. Van den Bosch, T.P.P., Kannegieter, N.M., Hesselink, D.A., Baan, C.C. & Rowhsani, A.T. (2017). Targeting the Monocyte-Macrophage Lineage in Solid Organ Transplantation. *Frontiers in Immunology* 8:153. https://doi: 10.3389/fimmu.2017.00153 (Accessed on 12 May 2021).

104. Wang, R., Wang, T. & Tait, S. (2017). Metabolic Regulation of Immunity. In: *Kelley and Firesteins's Textbook of Rheumatology (10th edn.) Vol. 1*; Firestein, G.S., Budd, R.C., Gabriel, S.E., McInnes, I.B. & O'Dell, J.R. (Eds.). Philadelphia: Elsevier, Inc. pp.318-326. https://doi.org/10.1016/B978-0-323-31696-5.00020-6 (Accessed on 2 June 2021).

105. Weller, C. (2021). Mast Cells. *British Society for Immunology*. https://www.immunology.org/public-information/bitesized-immunology/cells/mast-cells (Accessed on 18 May 2021).

106. Whitsett, J. & Alenghat, T. (2015). Respiratory epithelial cells orchestrate pulmonary innate immunity. *Nature Immunology* 16: 27-35. https://doi.org/10.1038/ni.3045 (Accessed on 20 July 2021).

107. Xiao, T.S. (2017). Innate immunity and inflammation. *Cellular and Molecular Immunology* 14(1): 1-3. https://doi: 10.1038/cmi.2016.45 (Accessed on 30 August 2021).

108. Xiong, S. & Xu, W. (2014). Human Immune System. *Comprehensive Biomedical Physics* 10: 91-114. https://doi.org/10.1016/B978-0-444-53632-7.01008-X (Accessed on 17 May 2021).

109. Yong, H.Y. & Luo, D. (2018). RIG-I-Like Receptors as Novel Targets for Pan-Antivirals and Vaccine Adjuvants Against Emerging and Re-Emerging Viral Infections. *Frontiers in Immunology* 9:1379. https://doi: 10.3389/fimmu.2018.01379 (Accessed on 16 June 2021).

110. Zaru, R. (2021). Pattern recognition receptor (PRRs) ligands. *British Society for Immunology*. https://www.immunology.org/public-information/bitesized-immunology/receptors-and-molecules/pattern-recognition-receptor-prrs (Accessed on 2 June 2021).

111. Zhang, J.-M. & An, J. (2007). Cytokines, Inflammation, and Pain. *International Anesthesiology Clinic* 45(2): 27-37. https://doi: 10.1097/AIA.0b013e318034194e (Accessed on 18 May 2021).

112. Zhao, C. & Zhao, W. (2020). NLRP3 Inflammasome-A Key Player in Antiviral Responses. *Frontiers in Immunology* 11:211. https://doi: 10.3389/fimmu.2020.00211/ (Accessed on 2 June 2021).

113. Zhong, Y., Kinio, A. & Saleh, M. (2013). Functions of NOD-like receptors in human diseases. *Frontiers in Immunology* 4:333. https://doi: 10.3389/fimmu.2013.00333 (Accessed on 11 June 2021).

FIGURE REFERENCES

114. Opening image. Wetzel, B. & Schaefer, H. (1982). Public domain; https://commons.wikimedia.org/wiki/File:SEM_blood_cells.jpg (Accessed on 26 April 2021).

115. Figure 2.1. (a) Beards, G. (2012). Creative commons attribution-sharealike 3.0 unported license; https://commons.wikimedia.org/wiki/File:Neutrophils.jpg (Accessed on 27 April 2021). (b) Lázaro-Díez, M., Chapartegui-González, I., Redondo-Salvo, S., Leigh, C., Merino, D., San Segundo, D., Fernández, A., Navas, J., Icardo, J.M., Acosta, F., Ocampo-Sosa, A., Martínez-Martínez, L. & Ramos-Vivas, J. (2017). Human Neutrophils phagocytose and kill Acinentobacter baumannii and A. pittii. Scientific Reports 10:4797. Creative commons attribution 4.0 international license; https://doi.org/10.1038/s41598-017-04870-8 (Accessed on 27 April 2021).

116. Figure 2.2. (Left) Rad, A., Häggström, M., Spacebirdy, RexxS, domdomegg (2016). Creative commons attribution-sharealike 4.0 international license, https://commons.wikimedia.org/wiki/File:Basophil.svg (Accessed on 3 May 2021). (Right) Blaus, B. & Blausen.com staff (2014). Medical gallery of Blausen Medical 2014. WikiJournal of Medicine 1(2). Creative commons attribution-sharealike 4.0 international license; https://doi:10.15347/wjm/2014.010. ISSN 2002-4436 (Accessed on 3 May 2021).

117. Figure 2.3. (Left) Uthman, E. (2010). Creative commons attribution 3.0 unported license; https://commons.wikimedia.org/wiki/File:Eosinophils_in_peripheral_blood.jpg (Accessed on 4 May 2021). (Right) Blaus, B. & Blausen.com staff (2014). Medical gallery of Blausen Medical 2014. WikiJournal of Medicine 1(2). Creative commons attribution-sharealike 4.0 international license; https://

doi:10.15347/wjm/2014.010. ISSN 2002-4436 (Accessed on 4 May 2021).

118. Figure 2.4. (a) Kauczuk (2006). Creative commons attribution license; https://commons.wikimedia.org/wiki/File:SMCpolyhydroxysmall.jpg (Accessed on 18 May 2021). (b) Krystel-Whittermore, M., Dileepan, K.N. & Wood, J.G. (2016). Mast Cell: a Multi-Functional Master Cell. Frontiers in Immunology 6:620. Creative commons attribution 4.0 international license; https://doi: 10.3389/fimmu.2015.00620 (Accessed on 18 May 2021).
119. Figure 2.5. (a) Grandjean, N. (2004). Creative commons attribution-sharealike 3.0 unported license; https://commons.wikimedia.org/wiki/File:Lymphocyte2.jpg (Accessed on 10 May 2021). (b) SpicyMilkBoy (2019). Creative commons attribution-sharealike 4.0 international license, https://commons.wikimedia.org/wiki/File:Reactive_lymphocyte.jpg (Accessed on 10 May 2021).
120. Figure 2.6. (Left) Beards, G. (2012). Creative commons attribution-sharealike 3.0 unported license, https://commons.wikimedia.org/wiki/File:Monocytes,_a_type_of_white_blood_cell_(Giemsa_stained).jpg (Accessed on 11 May 2021). (Right) Blaus, B. & Blausen.com staff (2014). Medical gallery of Blausen Medical 2014. WikiJournal of Medicine 1(2). Creative commons attribution-sharealike 4.0 international license; https://doi:10.15347/wjm/2014.010. ISSN 2002-4436 (Accessed on 11 May 2021).
121. Figure 2.7. van den Bosch, T.P.P., Kannegieter, N.M., Hesselink, D.A., Baan, C.C. & Rowhsani, A.T. (2017). Targeting the Monocyte-Macrophage Lineage in Solid Organ Transplantation. Frontiers in Immunology 8:153. Creative commons attribution 4.0 international license; https://doi: 10.3389/fimmu.2017.00153 (Accessed on 12 May 2021).
122. Figure 2.8. Coelho, C. (2013). Creative commons attribution-sharealike 4.0 international license; https://commons.wikimedia.org/wiki/File:Macrophages_undergo_mitosis_after_ingesting_a_fungal_cell.jpg (Accessed on 17 May 2021).
123. Figure 2.9. De Kleer, I., Willems, F., Lambrecht, B. & Goriely, S. (2014). Ontogeny of myeloid cells. Frontiers in Immunology 5:423. Creative commons attribution 3.0 unported license; https://doi: 10.3389/fimmu.2014.00423 (Accessed on 14 May 2021).

124. Figure 2.10. Askazal (2008). Creative commons attribution-sharealike 3.0 unported license; https://commons.wikimedia.org/wiki/File:Follicular_Dendritic_Cell.jpg (Accessed on 17 May 2021).

125. Figure 2.11. Castell-Rodríguez, A., Piñón-Zárate, G., Herrera-Enríquez, M., Jarquín-Yáñez, K. & Medina-Solares, I. (2017). Dendritic Cells: Location, Function, and Clinical Implications. In: Biology of Myelomonocytic Cells; Ghosh, A. (Ed.). Intechopen. Creative commons attribution 3.0 license; https://doi: 10.5772/intechopen.68352 (Accessed on 14 May 2021).

126. Figure 2.12. Castell-Rodríguez, A., Piñón-Zárate, G., Herrera-Enríquez, M., Jarquín-Yáñez, K. & Medina-Solares, I. (2017). Dendritic Cells: Location, Function, and Clinical Implications. In: Biology of Myelomonocytic Cells; Ghosh, A. (Ed.). Intechopen. Creative commons attribution 3.0 license; https://doi: 10.5772/intechopen.68352 (Accessed on 14 May 2021).

127. Figure 2.13. Proudfoot, A.E.I., Johnson, Z., Bonvin, P. & Handel, T.M. (2017). Glycosaminoglycan Interactions with Chemokines Add Complexity to a Complex System. Pharmaceuticals 10(3): 70. Creative commons attribution 4.0 international license; https://doi.org/10.3390/ph10030070 (Accessed on 31 May 2021).

128. Figure 2.14. Kopitar-Jerala, N. (2017). The Role of Interferons in Inflammation and Inflammasome Activation. Frontiers in Immunology 8: 873. Creative commons attribution 4.0 international license; https://doi: 10.3389/fimmu.2017.00873 (Accessed on 31 May 2021).

129. Figure 2.15. Salomon, B.L., Leclerc, M., Tosello, J., Ronin, E., Piaggio, E. & Cohen, J.L. (2018). Tumor Necrosis Factor α and Regulatory T Cells in Oncoimmunology. Frontiers in Immunology 9:444. Creative commons attribution 4.0 international license; https://doi: 10.3389/fimmu.2018.00444 (Accessed on 1 June 2021).

130. Figure 2.16. Immcarle105 (2020). Creative commons attribution-sharealike 4.0 international license; https://commons.wikimedia.org/wiki/File:PAMPs_and_PRRs_in_the_Innate_Immune_System.png (Accessed on 7 June 2021).

131. Figure 2.17. Gao, W., Xiong, Y., Li, Q. & Yang, H. (2017). Inhibition of Toll-Like Receptor Signaling as a Promising Therapy for Inflammatory Diseases: A Journey from Molecular to Nano Therapeutics. Frontiers in Physiology 8:508. Creative commons attribution 4.0 international

license; https://doi: 10.3389/fphys.2017.00508 (Accessed on 7 June 2021).

132. Figure 2.18. Jang, J.-H., Shin, H.W., Lee, J.M., Lee, H.-W., Kim, E.-C. & Park, S.H. (2015). An Overview of Pathogen Recognition Receptors for Innate Immunity in Dental Pulp. Mediators of Inflammation 794143. Creative commons attribution 3.0 unported license;https://doi.org/10.1155/2015/794143 (Accessed on 7 June 2021).
133. Figure 2.19. Zhong, Y., Kinio, A. & Saleh, M. (2013). Functions of NOD-like receptors in human diseases. Frontiers in Immunology 4:333. Creative commons attribution 3.0 unported license; https://doi: 10.3389/fimmu.2013.00333 (Accessed on 11 June 2021).
134. Figure 2.20. Zhong, Y., Kinio, A. & Saleh, M. (2013). Functions of NOD-like receptors in human diseases. Frontiers in Immunology 4:333. Creative commons attribution 3.0 unported license; https://doi: 10.3389/fimmu.2013.00333 (Accessed on 11 June 2021).
135. Figure 2.21. Zhong, Y., Kinio, A. & Saleh, M. (2013). Functions of NOD-like receptors in human diseases. Frontiers in Immunology 4:333. Creative commons attribution 3.0 unported license; https://doi: 10.3389/fimmu.2013.00333 (Accessed on 11 June 2021).
136. Figure 2.22. Zhao, C. & Zhao, W. (2020). NLRP3 Inflammasome-A Key Player in Antiviral Responses. Frontiers in Immunology 11:211. Creative commons attribution 4.0 international; https://doi: 10.3389/fimmu.2020.00211/ (Accessed on 2 June 2021).
137. Figure 2.23. Brisse, M. & Ly, H. (2019). Comparative Structure and Function Analysis of the RIG-I-Like Receptors: RIG-I and MDA5. Frontiers in Immunology 10:1586. Creative commons attribution 4.o international license; https://doi: 10.3389/fimmu.2019.01586 (Accessed on 16 June 2021).
138. Figure 2.24. Yong, H.Y. & Luo, D. (2018). RIG-I-Like Receptors as Novel Targets for Pan-Antivirals and Vaccine Adjuvants Against Emerging and Re-Emerging Viral Infections. Frontiers in Immunology 9:1379. Creative commons attribution 4.0 international license;https://doi: 10.3389/fimmu.2018.01379 (Accessed on 16 June 2021).
139. Figure 2.25. Kerrigan, A.M. & Brown, G.D. (2009). C-type lectins and phagocytosis. Immunobiology 214(7): 562-575. Creative commons attribution 3.0 unported license; https://doi.org/10.1016/j.imbio.2008.11.003 (Accessed on 22 June 2021).

140. Figure 2.26. Chiffoleau, E. (2018). C-Type Lectin-Like Receptors As Emerging Orchestrators of Sterile Inflammation Represent Potential Therapeutic Targets. Frontiers in Immunology 9:227. Creative commons attribution 4.0 international license; https://doi: 10.3389/fimmu.2018.00227 (Accessed on 22 June 2021).
141. Figure 2.27. Drouin, M., Saenz, J. & Chiffoleau, E. (2020). C-Type Lectin-Like Receptors: Head or Tail in Cell Death Immunity. Frontiers in Immunology 11:251. Creative commons attribution 4.0 international license;https://doi: 10.3389/fimmu.2020.00251 (Accessed on 22 June 2021).
142. Figure 2.28. Biga, L.M., Dawson, S., Harwell, A., Hopkins, R., Kaufmann, J., LeMaster, M., Matern, P., Morrison-Graham, K., Quick, D. & Runyeon, J. (2021). Anatomy & Physiology. Creative commons attribution-sharealike 4.0 international license; https://open.oregonstate.education/aandp/chapter/21-2-barrier-defenses-and-the-innate-immune-response/ (Accessed on 26 April 2021).
143. Figure 2.29. Abdallah, F., Mijouin, L. & Pichon, C. (2017). Skin Immune Landscape: Inside and Outside the Organism. Mediators of Inflammation 5095293. Creative commons attribution 4.0 international license; https://doi.org/10.1155/2017/5095293 (Accessed on 30 June 2021).
144. Figure 2.30. Coates, M., Blanchard, S. & MacLeod, A.S. (2018). Innate antimicrobial immunity in the skin: a protective barrier against bacteria, viruses, and fungi. PLoS Pathogens 14(12): e1007353. Creative commons attribution 4.0 international license; https://doi.org/10.1371/journal.ppat.1007353 (Accessed on 24 June 2021).
145. Figure 2.31. Abdallah, F., Mijouin, L. & Pichon, C. (2017). Skin Immune Landscape: Inside and Outside the Organism. Mediators of Inflammation 5095293. Creative commons attribution 4.0 international license; https://doi.org/10.1155/2017/5095293 (Accessed on 30 June 2021).
146. Figure 2.32. Parigi, S.M., Eldh, M., Larssen, P., Gabrielsson, S. & Villalanca, E.J. (2015). Breast milk and solid food shaping intestinal immunity. Frontiers in Immunology 6: 415. Creative commons attribution 4.0 international license; https://doi: 10.3389/fimmu.2015.00415 (Accessed on 13 July 2021).
147. Figure 2.33. McGhee, J.R. & Fujihashi, K. (2012). Inside the Mucosal Immune System. PLoS Biology 10(9): e1001397. Creative commons

attribution license; https://doi.org/10.1371/journal.pbio.1001397 (Accessed on 19 July, 2021).

148. Figure 2.34. Silva, F.A.R., Rodrigues, B.L., Ayrizono, M.d.L.S. & Leal, R.F. (2016). The Immunological Basis of Inflammatory Bowel Disease. Gastroenterology Research and Practice 2097274. Creative commons attribution 4.0 international license;https://doi.org/10.1155/2016/2097274 (Accessed on 14 July 2021).
149. Figure 2.35. Frey, A., Lunding, L.P., Ehlers, J.C., Weckmann, M., Zissler, U.M. & Wegmann, M. (2020). More Than Just a Barrier: The Immune Functions of the Airway Epithelium in Asthma Pathogenesis. Frontiers in Immunology 11: 761. Creative commons attribution 4.0 international license; https://doi: 10.3389/fimmu.2020.00761 (Accessed on 3 August 2021).
150. Figure 2.36. Kumar, V. (2020). Pulmonary Innate Immune Response Determines the Outcome of Inflammation During Pneumonia and Sepsis-Associated Acute Lung Injury. Frontiers in Immunology 11: 1722. Creative commons attribution 4.0 international license, https://doi: 10.3389/fimmu.2020.01722 (Accessed on 2 August 2021).
151. Figure 2.37. Uribe-Querol, E. & Rosales, C. (2020). Phagocytosis: Our Current Understanding of a Universal Biological Process. Frontiers in Immunology 11:1066. Creative commons attribution 4.0 international license; https://doi: 10.3389/fimmu.2020.01066 (Accessed on 10 August 2021).
152. Figure 2.38. Uribe-Querol, E. & Rosales, C. (2020). Phagocytosis: Our Current Understanding of a Universal Biological Process. Frontiers in Immunology 11:1066. Creative commons attribution 4.0 international license; https://doi: 10.3389/fimmu.2020.01066 (Accessed on 10 August 2021).
153. Figure 2.39. Uribe-Querol, E. & Rosales, C. (2020). Phagocytosis: Our Current Understanding of a Universal Biological Process. Frontiers in Immunology 11:1066. Creative commons attribution 4.0 international license; https://doi: 10.3389/fimmu.2020.01066 (Accessed on 10 August 2021).
154. Figure 2.40. Uribe-Querol, E. & Rosales, C. (2020). Phagocytosis: Our Current Understanding of a Universal Biological Process. Frontiers in Immunology 11:1066. Creative commons attribution 4.0 international license; https://doi: 10.3389/fimmu.2020.01066 (Accessed on 10 August 2021).

155. Figure 2.41. Uribe-Querol, E. & Rosales, C. (2020). Phagocytosis: Our Current Understanding of a Universal Biological Process. Frontiers in Immunology 11:1066. Creative commons attribution 4.0 international license; https://doi: 10.3389/fimmu.2020.01066 (Accessed on 10 August 2021).

156. Figure 2.42. Uribe-Querol, E. & Rosales, C. (2020). Phagocytosis: Our Current Understanding of a Universal Biological Process. Frontiers in Immunology 11:1066. Creative commons attribution 4.0 international license; https://doi: 10.3389/fimmu.2020.01066 (Accessed on 10 August 2021).

157. Figure 2.43. Biga, L.M., Dawson, S., Harwell, A., Hopkins, R., Kaufmann, J., LeMAster, M., Matern, P., Morrison-Graham, K., Quick, D. & Runyeon, J. (2021). Anatomy & Physiology (1st edn.). Oregon State University. Creative commons attribution-sharealike 4.0 international license. https://open.oregonstate.education/aandp/chapter/21-2-barrier-defenses-and-the-innate-immune-response/ (Accessed on 8 September 2021).

CHAPTER 3

ADAPTIVE IMMUNE SYSTEM

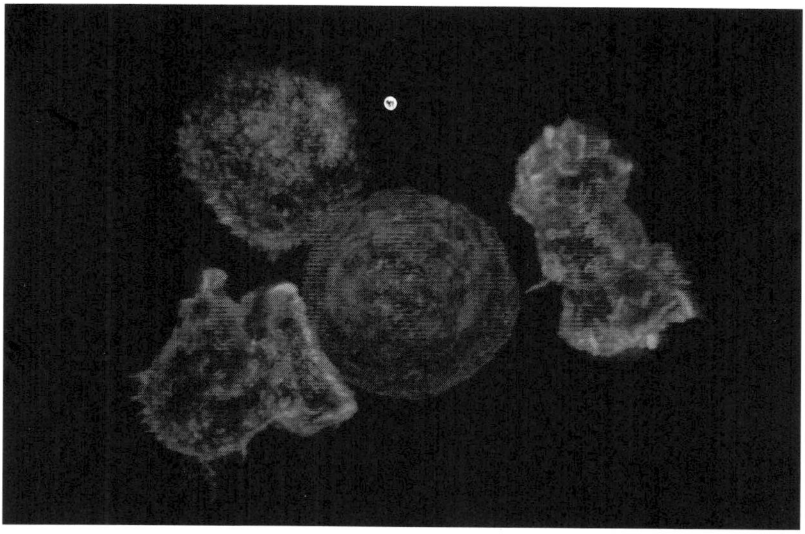

(Source: National Institutes of Health, Public Domain)

A critical function of the innate immune system is to recruit a more sophisticated, pathogen-specific immune response via the adaptive immune system, which eliminates pathogens that evaded the innate defenses. There are two types of adaptive immunity, the humoral or antibody response and the cell-mediated response. Any substance that can induce an adaptive immune response is known as an antigen (Ag), which stands for *anti*body *gen*erator (Alberts *et al.*, 2002). There are self- and non-self-antigens, which can be distinguished by certain receptors so that foreign antigens are eliminated while self-antigens are left untouched. Unlike the immediate innate immune response, this finely tuned immune system takes a while to be activated requiring about 3-5 days. This indicates that adaptive immunity has several steps, which include recognition of the antigen, the interaction of different lymphocyte subsets, activation and proliferation of the responding cells, gene expression, and production of certain end products like antibodies or immunoglobulins and cytokines. After the clearance of the pathogen, memory immune cells are generated that can provide long-term, quicker, and amplified protection upon re-exposure to certain microbes (Mirzaei, 2020; Moticka, 2016).

3.1 CELLS OF THE ADAPTIVE IMMUNE SYSTEM

Some of the cells of the innate immune system, which are described in the previous chapter, may also have certain roles in the adaptive immune response. The main cells of the adaptive immune system, however, are the lymphocytes.

3.1.1 Lymphocytes

Lymphocytes are agranular leukocytes that originate from hematopoietic stem cells in the bone marrow. The immature ones are relatively smaller with their nuclei occupying most of the cellular space. When activated by antigens, they enlarge so that their cytoplasmic content and organelle number increase. Lymphocytes are the only cells that can recognize and respond specifically to each antigen (Britannica, 2021; Cano & Lopera, 2013).

3.1.1.1. T lymphocytes

Some of the immature lymphocytes are released from the bone marrow into the bloodstream and then travel to the thymus where they multiply and differentiate into the so-called T lymphocytes or T cells (i.e., thymus-derived). The T cells enter the blood circulation and move to and within the rest of the lymphoid organs, where they multiply further in response to certain stimuli. Nearly half of all lymphocytes are T cells. These are the principal cells involved in cell-mediated immune response, i.e., they recognize only antigens that have entered the cells like viruses. They are also critical in regulating the function of B cells (Britannica, 2021).

As the T cells mature, they acquire a T cell receptor (TCR) that recognizes a specific antigen. The TCR does not recognize the antigen in its natural form. The antigen needs to be digested or processed into a linear peptide and presented by antigen presenting cells(APCs) via major histocompatibility complex molecules (MHCs), also referred to as the human leukocyte antigens (HLAs) since these were first discovered on leukocytes (see Figure 3.1 and 3.2). T cells also have either CD4 or CD8 which are cell surface proteins that serve as co-receptors strengthening the binding between the T cell and the target cell (Cano & Lopera, 2013; Britannica, 2021).

MHCs are expressed on the surface of APCs (see Figure 3.1 and 3.2). These are classified as either class I (also known as HLA I with the following members: HLA-A, HLA-B, and HLA-C) that are found on all nucleated cells or class II (also called HLA II with members like HLA-DP, HLA-DQ, and HLA-DR) that occur on certain immune cells like macrophages, dendritic cells (DCs), B cells and some T cells. MHC I presents intracellular peptides with 8-10 amino acids to T cells. These peptides are from proteins translated within the cell that are encoded by the host genome or by the genome of a pathogen that infected the cell. Meanwhile, MHC II presents extracellular peptides with 13-25 amino acids. These peptides are from proteins that were endocytosed and then released from the endosome to the cytoplasm so that they are digested by proteasome. The resulting peptides are then transported by the transporter associated with antigen processing(TAP) into the ER where they associate with MHC molecule and then transported to the cell surface to be detected by T cells. There is a high frequency for T cells to interact with an APC with the appropriate MHC-peptide complex since T cells are circulating throughout the body in the lymphatic and circulatory system and are accumulated along with APCs in lymph nodes (Warrington *et al*., 2011; Bhagavan & Ha, 2011; Cano & Lopera, 2013; Bonilla & Oettgen, 2010).

T cells mature and differentiate into either helper T (Th) cells and cytotoxic T cells. When the T cell expresses the CD4 and it is activated by interacting with peptides bound to MHC II on APCs, it turns into a Th cell. Th cells do not directly attack other cells but produce cytokines and stimulate B cells to produce immunoglobulins as well as activate cytotoxic T cells and macrophages to attack infected cells (see Figure 3.1). Depending on the cytokines they produce, there are several distinct phenotypes of Th cells, including Th1, Th2, Th9, Th17, Th22, regulatory T (Treg) cells, and follicular helper T (Tfh) cells (Cano & Lopera, 2013; Britannica, 2021). Th cells are cleared by phagocytes once the infection has been cleared with a few remaining as Th memory cells (Warrington *et al*., 2011).

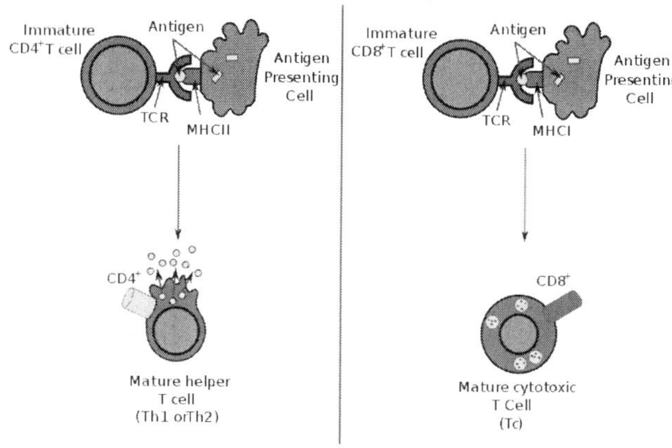

Figure 3.1. A T cell that expresses CD4 can be activated by antigen presenting cells (APCs) that present the ingested antigen via a major histocompatibility complex (MHC) molecule to the T cell receptor (TCR). This leads to the conversion of the T cell into a helper Tcell. (Source: Sjef, Creative Commons License)

T cells may also express CD8 and this type of T cells may be activated by peptides bound to MHC I to develop into cytotoxic T cells that directly attack and destroy malignant or virus-infected cells(see Figure 3.2). They induce apoptosis by releasing cytolytic granules or by expressing FasL, a ligand for death receptors (Cano & Lopera, 2013; Marshall *et al.*, 2018).

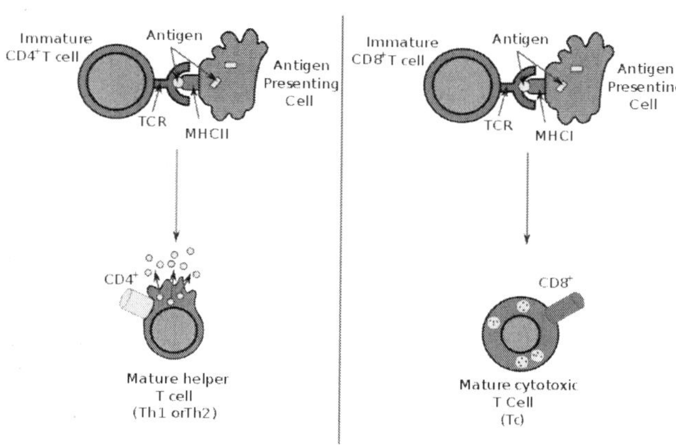

Figure 3.2. A T cell expressing CD8 may be activated by antigens presented by major histocompatibility complex class I (MHC I) on an antigen presenting cell (APC) so that it matures into a cytotoxic T cell. (Sjef, Creative Commons License)

The random generation of different TCRs produces a large proportion of T cells that recognize self-antigens and can attack the body's own tissues. These self-reactive T cells die before they leave the thymus so that only those that can recognize foreign antigens remain and are released into circulation (Britannica, 2021).

3.1.1.2. B lymphocytes

Other lymphocytes remain in the bone marrow, where they differentiate into the so-called B lymphocytes or B cells. Their name was derived from the lymphoid organ in birds known as bursa of Fabricius, where these cells were first discovered. Like the T cells, they also move to the lymphoid organs like the spleen, lymph nodules, Peyer's patches, tonsils, and mucosal tissues, where they mature into transitional B cells via distinct cell stages called T1 and T2 (see Figure 3.3) (Britannica, 2021; Cano & Lopera, 2013; Roghanian & Newman, 2021).

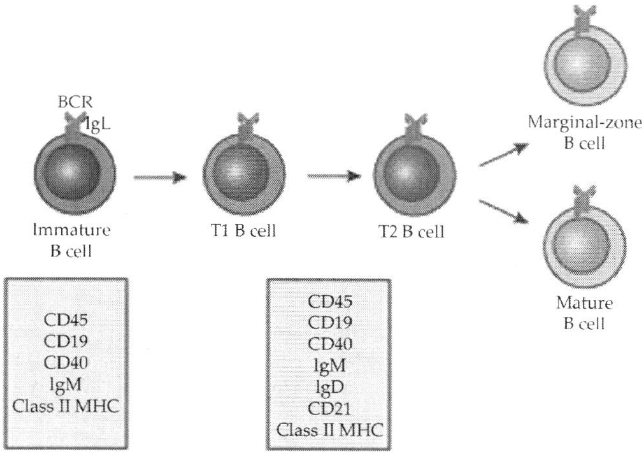

Figure 3.3. Immature B cell leaves the bone marrow and migrates to lymphoid organs, such as the spleen, where it matures into a transitional B cell (T1 and T2) that further differentiates into either a marginal zone B cell or a mature (follicular) B cell. (Source: Bobologist, Creative Commons License)

Eventually, those located in the marginal zones of the spleen differentiate into marginal zone B cells (see Figure 3.3). These also reside in the inner wall of the subcapsular sinus of the lymph nodes, the epithelium of tonsillar crypts, and the subepithelial dome of intestinal Peyer's patches. They express high levels of TLRs allowing them to mediate between innate and adaptive immune responses. Other transitional B cells enter the germinal centers of secondary lymphoid organs and become follicular B cells. These are the resting or naïve cells that compose the largest subpopulation of B cells. The movement of the B cells to the marginal zone or the follicles leads to the formation of germinal centers, a specialized microenvironment of lymphoid tissue where B cell proliferation, somatic hypermutation, and selection by antigenic affinity occur (Cano & Lopera, 2013). In the germinal centers, the B cells are then committed to the synthesis and expression of a single type of B cell receptor (BCR).

The first B cells to form in the fetal liver are the B1 B cells. The initial wave of lymphopoiesis in the embryo favors the formation of this B cell type. They occur in peritoneal and pleural cavities as well as in the spleen and intestine. They are the main source of circulating immunoglobulins. Meanwhile, mature peripheral B cells are found in tonsils (Cano & Lopera, 2013; Roghanian & Newman, 2021).

Finally, antigenic and cytokine stimulation trigger the proliferation and transformation of each of these cells into plasma cells or memory B cells(see Figure 3.4). Plasma cells are considered to be effector cells that defend the body by secreting immunoglobulins. They no longer express surface molecules like CD19, CD20, CD22, MHC II, and BCR. They also lose their ability to divide. On the other hand, their endoplasmic reticulum significantly grows and harbors more ribosomes resulting to the synthesis of more immunoglobulins as well as an increase in their cytoplasmic area. Plasma cells may be short-lived or long-lived. Short-lived plasma cells do not have somatically mutated immunoglobulin genes and they provide a quick initial immune response. They reside in the medulla of the ganglia and then quickly enter the bloodstream and find the antigen to initiate the synthesis of specific IgM. Long-lived plasma cells have high affinity BCRs produced by several rounds of proliferation and affinity maturation. They have a prolonged survival due to the effect of IL16. They move to a special niche in the bone marrow, where IgGs are produced to mount a prolonged or permanent defense against the antigen that originally activated the B cells. Memory B cells are long-lasting, and they are responsible for the lifetime immunities to certain childhood diseases. There are also different classes of memory B cells depending on their origin, i.e., the spleen, the germinal center, and the intestine lamina propria (Cano & Lopera, 2013; Roghanian & Newman, 2021).

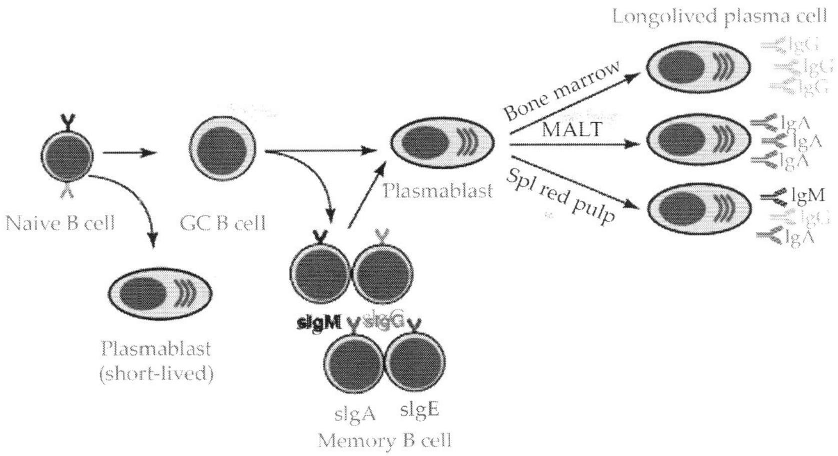

Figure 3.4. A naïve B cell may be stimulated by antigenic exposure and cytokines to develop into either short-lived plasma cell (plasmablast), long-lived plasma cell or memory cell. (Source: Bobologist, Creative Commons License)

Another subset of B cells is the regulatory B (Breg) cells, which play a key role in maintaining immunotolerance. Breg cells produce IL10, which inhibits pro-inflammatory cytokines, reduces the expression of MHC II molecules, and facilitates the differentiation of Treg cells. These result in the suppression of excessive inflammatory response during autoimmune diseases or that can be caused by unresolved infections (Cano & Lopera, 2013; Roghanian & Newman, 2021).

3.1.2 B Cell Helpers

Apart from Th cells, there are other cells that assist B cells, which include some innate immune cells (i.e., DCs, macrophages, neutrophils, eosinophils, basophils, and mast cells). Other specialized B cell helpers include invariant natural killer T (iNKT) cells and B cell helper neutrophils (N_{BH}).

The iNKT cells express a TCR variant that recognizes soluble glycolipids presented by DCs or macrophages. These cells interact with Tfh cells to form active germinal centers resulting in the subsequent production of long-lived plasma cells that secrete IgG (Cano & Lopera, 2013).

Under healthy conditions, the N_{BH} cells are found in the perimarginal zone of the spleen. They are stimulated by cytokines and microbial products to interact with perifollicular B cells and marginal zone B cells. This interaction leads to class switch recombination wherein plasma cells switch from producing IgM to secreting IgG and IgA (Cano & Lopera, 2013).

3.2 ANTIBODY IMMUNE RESPONSE

B cells are activated to produce proteins called antibodies or immunoglobulins, which circulate in the blood and permeate the other body fluids, where they bind to a specific antigen or antigenic determinants, called epitopes, (e.g., proteins or glycoproteins, nucleic acids, carbohydrates, and lipids) that triggered their secretion (Aktor, 2012). Antigens are generally high molecular weight proteins or polysaccharides. Polypeptides, lipids, nucleic acids, and other types of compounds may also function as antigens. These are usually from a foreign source (e.g., viruses, bacteria, protozoans, fungi, pollen, dust, transplanted tissue) that enter the host via an infection. However, there are cases when the body's own proteins may act as antigens and trigger an autoimmune response (Merck KGaA, 2021; Britannica, 2021). Antigens that trigger an immune response are known as immunogens, whereas those that can bind a lymphocyte receptor but do not induce an immune response are called haptens. Haptens can become immunogenic when bound to a larger molecule like a protein (Britannica, 2021).

Antigens are bound to the immunoglobulins at the hypervariable regions via multiple forces, which include electrostatic interactions, hydrogen binding, hydrophobic interactions, and van der Waals forces (Aktor, 2012). When viruses and microbial toxins are bound by immunoglobulins, they are prevented from binding receptors on host cells. Therefore, they cannot enter the host cells and are said to be neutralized. In addition, antibodies bind pathogens to mark them for phagocytosis so that they are prevented from replicating outside the cells (Alberts *et al*., 2002). Antibodies bound to the pathogen surface can also activate the complement system resulting in the opsonization of pathogens so that these are phagocytosed or lysed (Janeway *et al*., 2001).

Antigen triggers the activation of B cells into antibody-secreting plasma cells. B cell activation also requires Th cells, which often refer to Th2 type of CD4 T cells, but in some cases, a subset of Th1 cells can help in B cell activation (Janeway *et al*., 2001).

3.2.1 Immunoglobulins

Immunoglobulins (Ig) or antibodies are glycoproteins that are produced by plasma cells. They constitute approximately 20% of the blood plasma proteins. When a certain antigen binds to a B cell receptor (BCR)on the surface of B lymphocytes, it activates a signaling cascade that induces transcription factors to initiate the expression of immunoglobulins that are specific to the antigen that stimulated the B cell. At this point, the B cell has differentiated to become a plasma cell (Justiz Vaillant *et al.*, 2021). Immunoglobulins and BCRs from one B cell are identical in terms of their antigen binding sites, except that BCRs have an extra domain that penetrates the cell membrane in order to anchor them to the plasma membrane. Therefore, immunoglobulins and BCRs have similar structures and properties (Britannica, 2021). IgM and IgD are the immunoglobulin classes with membrane-bound forms that become BCRs.

Immunoglobulins consist of four polypeptides, i.e., two light chains that are approximately 25 kiloDaltons (kDa) each and two heavy chains, each of which is about 50 kDa (see Figure 3.5). The heavy chains differ among the different types of immunoglobulins, whereas the light chain can be made up of either a κ or a λ chain. The two heavy chains are linked together by disulfide bonds and each heavy chain is linked by another disulfide bridge to a light chain to form a Y-shaped molecule. The amino terminal regions of the chains, which constitute the arms of the Y, fold to form the variable (V) domains, so-named because they vary in amino acid sequence among different immunoglobulins (Janeway *et al.*, 2001). In the V domain of each light and heavy chain, there are three hypervariable regions (HRRs), also known as complementarity determining regions (CDRs). These fold into regions that produce two antigen-binding sites at the tip of each monomer so that an immunoglobulin can bind to two identical antigens simultaneously (Justiz Vaillant *et al.*, 2021). The carboxy termini have the constant (C) domains with the same or fewer variations in amino acid sequence, and these form the stem of the Y. The heavy chain has three or four C domains, while the light chain contains only one. Heavy chains with three C domains may include a spacer hinge region between the first and the second C domains (Schroeder& Cavacini, 2010).

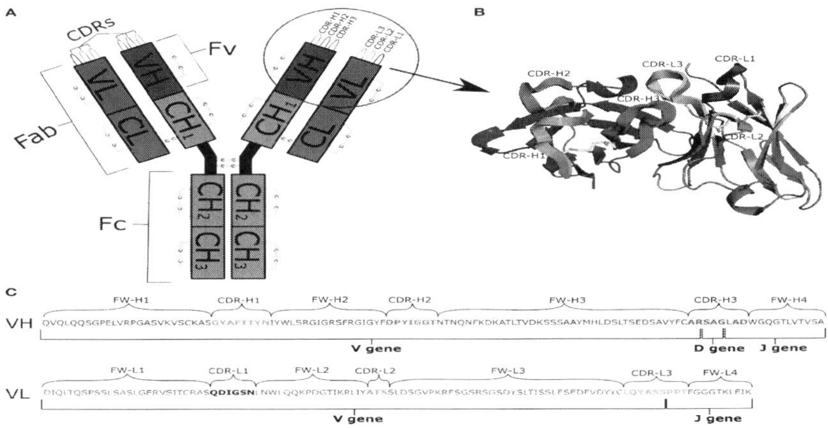

Figure 3.5. (A) An immunoglobulin is made up of two light chains and two heavy chains having a Y-like conformation. The heavy chains are linked by a disulfide bond (yellow) between two cysteine residues (C-C) at the hinge region (black). Each light chain is also joined to the heavy chain via disulfide linkages. Each chain is composed of one variable (V) as well as one constant domain in the light chain (CL) and two to three constant domains in the heavy chain (CH). When an immunoglobulin is cleaved by a protease, it produces the crystallizable fragment (Fc) and two antigen binding fragments (Fab), which, in turn, can be further split into the variable (Fv) and constant regions. (B) Fv is made up of VH and VL domains, each of which has three complementarity determining regions (CDRs). (Source: Aleksandr Kovaltsuk *et al.*, Creative Commons License)

Digestion of the immunoglobulin with the protease papain produces one Fc (crystallizable)fragment and two Fab (antigen binding)fragments or a single dimeric Fab (see Figure 3.5). The Fc region that mediates biological functions, such as binding to cellular receptors and effector cells, consists of the constant domains made up of a pair of C_H2 that do not interact with each other and a pair of C_H3 domains that are joined together. On the other hand, the Fab region has the antigen binding sites, and it contains one complete light chain as well as the variable and first constant domain of one heavy chain. The Fab can be subdivided into a variable fragment (Fv) made up of the variable domains of the heavy (V_H) and light chains (V_L) and a constant fragment composed of the C_L and the C_H1 domains (Schroeder & Cavacini, 2010; Janeway *et al.*, 2001).

The immunoglobulin is a flexible molecule particularly at the hinge region, which allows independent movement of the two Fab arms. The junction between the V and C domains also enables bending and rotation of the V domain so that it is called a molecular ball and-socket joint.

The flexibility afforded by the hinge and socket allows both arms of the immunoglobulin to bind to sites that are various distances apart and enables the immunoglobulins to engage with the antibody-binding proteins that mediate immune effector processes (Janeway *et al*., 2001).

3.2.1.1. Different isotypes of immunoglobulins

The different classes or isotypes of immunoglobulins differ in structure resulting in their differences in function and antigen responses. There are five major classes in placental mammals based on differences in amino acid sequence in the Fcof the heavy chains. These are IgM, IgG, IgA, IgE, and IgD with IgG and IgA having subtypes. IgM has the mu (μ) heavy chain isotype, IgD has delta (δ), IgG has gamma (γ), IgA has alpha (α), and IgE has epsilon (ε). The distinguishing function of each type is conferred by the carboxy-terminal region of the heavy chain (Janeway *et al*., 2001).

Immunoglobulin M

IgM may exist as a monomer, pentamer or hexamer (see Figure 3.6). The pentameric IgM has a molecular weight of 970kDa, and its average concentration in the blood serum is 1.5 mg/ml. On the other hand, the hexameric form is more than 1,000 kDa in molecular weight. IgM is a potent **agglutinin**, a substance that causes particles to coagulate and aggregate(Vaillant *et al*., 2021; Jones *et al*., 2020).

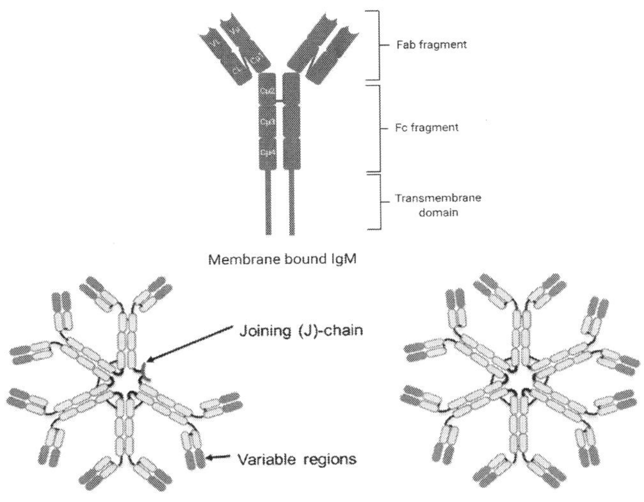

Figure 3.6. (Upper) The monomeric and membrane-bound IgM contains Fab (antigen-binding) and Fc (crystallizable) fragments as well as a transmembrane

signaling tail that binds to Fc receptors on the surface of B cells. (Lower left) The pentameric IgM has 10 antigen binding sites in its variable regions (green), and it is characterized by the presence of a joining chain (J-chain). This chain is absent in the (lower right) hexameric form. (Source: Upper image- Katelyn Jones *et* al., Creative Commons License; Lower images-Bruce A. Keyt *et al.*, Creative Commons License)

IgM maybe membrane-bound or secreted. Membrane-bound IgMis monomeric, and it can bind certain PAMPs such as particular carbohydrates and lipopolysaccharide. These are also called natural antibodies. During the primary response to infection, naïve B cells express monomeric IgM on their plasma membrane that functions as BCR associating with CD79a and CD79b, which are polypeptide chains that participate in IgM cell signaling. Mature and antigen-stimulated B cells secrete soluble pentameric IgM with subunits linked together by disulfide bonds in the C_H4 domain. The pentamer contains the joining or J-chain that facilitates secretion at mucosal surfaces (see Figure 3.6). It is the first antibody that is synthesized during B cell development so that it is more polyreactive than other isotypes. Since it is part of the first line of defense against invading microbes, it is part of the innate immune system. While the monomeric form has low affinity (i.e., the strength of a single interaction) to antigens, the multimeric interactions between the pentamer, via its 10 antigen binding sites, and the antigen results in high avidity (i.e., strength of all interactions that bind the antigen and antibody). Therefore, the pentameric IgM is very efficient in fixing the complement component C1q and opsonizing target particles for elimination (see Figure 3.7). The role of the hexameric IgM is currently unknown, but it may be formed due to defects in the μ chain or J-chain regions in the pentameric form (Schroeder & Cavacini, 2010; Keyt *et al.*, 2020; Jones *et al.*, 2020).

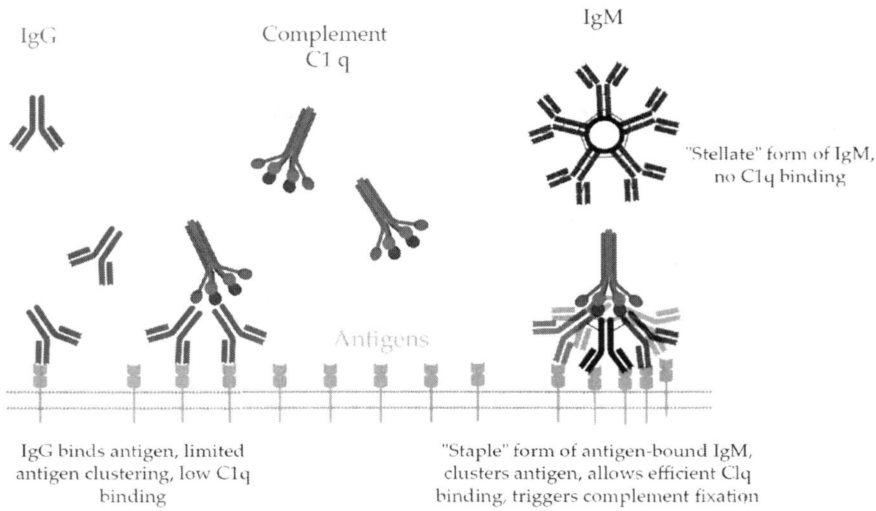

Figure 3.7. Both IgM and IgG can activate the complement system leading to complement-dependent cytotoxicity (CDC). IgM's hexameric or pentameric structure provides multiple antigen binding sites and enables highly avid binding to the complement component C1q. Thus, IgM can fix complement significantly better than IgG. (Source: Bruce A. Keyt *et al*., Creative Commons License)

Through its Fc domain, IgM can bind to multiple receptors, namely the complement receptors CR2 and CR3, the polymeric Ig receptor (pIgR) and two Fc receptors, the Fcα/μ receptor (Fcα/μR)and FcμR (see Figure 3.8). The pIgR is a highly glycosylated receptor expressed on the basolateral surfaces of mucosal epithelial cells in tissues like the lungs, kidneys, and endometrium, but with the highest expression in the small and large intestines. Its secretory component (SC) binds polymeric immunoglobulins with J-chain such as IgM and IgA, after which the immunoglobulin-pIgR complex is phagocytosed and transported via endosomes from the basal to the apical cell surface. A proteolytically sensitive region of the receptor, which is proximal to the plasma membrane, is cleaved so that the polymeric immunoglobulin is released from the cell (Keyt *et al*., 2020). Thus, pIgR mediates the transcytosis of secreted IgM and IgA from lamina propria of the epithelium to apical mucosal sites where the immunoglobulin is released into the secretions (Jones *et al*., 2020).

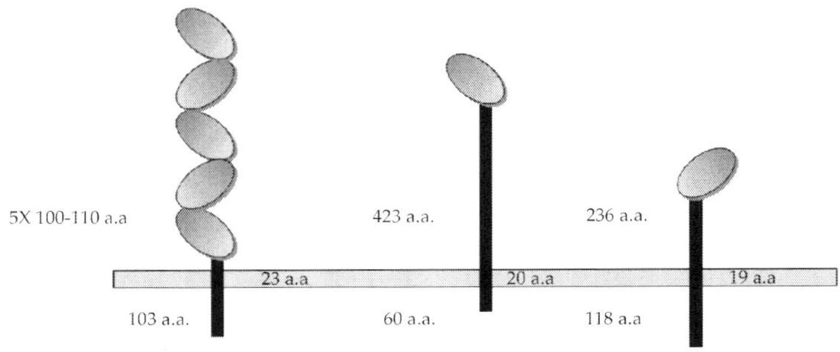

Receptor	pIgR	Fcα/μR	FcμR
Requires J-chain for binding IgM?	Yes	Yes	No
Expressed in	Small and large intestine, lungs, kidneys, mucosal epithelium	Lymph node, appendix, kidney, B-lymphocytes, macrophages	T-and B-lymphocytes

Figure 3.8. IgM can bind at least three types of receptors, namely the polymeric Ig receptor (pIgR), Fcα/μ receptor (Fcα/μR), and FcμR. They differ in terms of their tissue distribution, their requirement for binding IgM, and the number of their immunoglobulin fold-like regions (oval). (Source: Bruce A. Keyt *et al.*, Creative Commons License)

The Fcα/μRis also glycosylated like the pIgR. It is located in lymphoid tissues including the lymph nodes and the appendix as well as in non-lymphoid tissues, such as the kidney and intestine, with lower expression levels in the lungs, liver, and heart. Its amino acid residues 76-98 are homologous to the CDR1 region of pIgR, which is a conserved binding site for both receptors. Just like pIgR, it is internalized once it binds IgM. This process could facilitate the processing and presentation of antigens to Th cells. Finally, the FcμR, also known as the TOSO/Fas apoptotic inhibitory molecule 3 (FAIM3) or Fc fragment of IgM receptor (FCMR), is a transmembrane sialoglycoprotein that is highly expressed on chronic lymphocytic leukemia B cells and observed also in other B cells and in T cells. It has a high affinity to multimeric IgM, but a lower affinity to monomeric ones. Just like the other receptor types, it is also internalized upon binding IgM. But unlike the other two receptor types, it does not recognize IgA, does not require a J-chain for binding IgM, and it has a much shorter CDR1 with only five amino acids. Its function is still a subject of research, but studies on mice indicate that it may regulate autoantibody production and facilitate pro-inflammatory cytokine synthesis (Keyt *et al.*, 2020; Gong & Ruprecht, 2020; Jones *et al.*, 2020).

Immunoglobulin G

IgG, the predominant isotype found in the body, is a monomer that is approximately 146 kDa in molecular weight with an average serum concentration of 9.0 mg/ml. It is described to be divalent with two identical antigen-binding sites, which are composed of two light chains and two heavy chains that are linked by disulfide bonds. It is mostly synthesized during the secondary immune response. It also activates the classical complement pathway. IgG is the only antibody that can cross the placenta since its Fc binds the receptors on the placental surface. Therefore, it protects the fetus from infectious diseases, and it is the most abundant immunoglobulin in newborns (Vaillant *et al*., 2021).

It has four subclasses that are identical at the amino acid level by more than 90%. However, they differ from each other in terms of the structure and function mostly of the C_H1, C_H3, and hinge regions (see Figure 3.9). These differences affect the antibody flexibility and functional affinity, which mediate cooperative interactions with multivalent antigens. In particular, the C_H1 domain and hinge region control the flexibility of the Fab and Fv portions of the immunoglobulin (Schroeder & Cavacini, 2010).

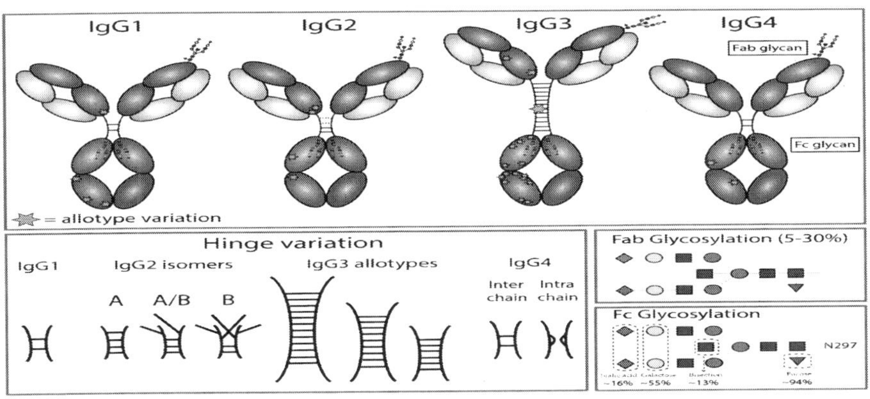

Figure 3.9. Immunoglobulin has four subclasses, namely IgG1, IgG2, IgG3, and IgG4. These differ in certain amino acid residues (small blue stars) in the constant domain of the heavy chains, in the hinge region (large blue star) as well as in their glycosylation. (Source: Steven W. de Taeye *et al*., Creative Commons License)

The four subtypes include IgG1, IgG2, IgG3, and IgG4 (see Figure 3.9). IgG3 is slightly larger than the other subtypes with a molecular weight of 165 kDa, and it is the first subtype to be formed. IgG1 is the next subtype to be produced eventually becoming the predominant one comprising about 61% of the total IgG, followed by IgG2, IgG3, and IgG4. IgG1 and IgG3 are generally triggered by protein antigens, whileIgG2 and IgG4are induced by polysaccharide antigens so that they can provide defense against encapsulated bacteria. Additionally, IgG1 and IgG3 are strong inducers of Fc-mediated effector mechanisms including complement-dependent cytotoxicity (CDC), antibody-dependent cellular cytotoxicity (ADCC), and antibody-dependent cellular phagocytosis (ADCP), while IgG2 and IgG4 induce subtle responses. It was also observed that IgG3 is more effective in neutralizing viruses than IgG1, which may be attributed to the greater flexibility of IgG3 enhancing its access or capacity to induce changes in the oligomer structure of the virus. Meanwhile, IgG4 is the only subclass that does not activate the complement pathway. It is often produced due to repeated or prolonged antigen exposure (Vaillant *et al*., 2021; Schroeder & Cavacini, 2010; ThermoFisher Scientific, 2021; de Taeye *et al*., 2019).

The four subclasses also differ in terms of their affinity to various subtypes of type I Fc receptors, FcγRs, i.e., FcγRI, FcγRII, and FcγRIII (see Table 2.3). Such differences do not always reflect the variation in their functionalities because it is the cross-linking of the Fc receptors, resulting from their engagement with immune complexes or opsonized pathogens with multiple IgG molecules, which initiates signaling for certain processes, such as ADCC of the target cell. FcγRs bind the Fc domain of IgG via their second extracellular domains. The low-affinity FcγRII (IIa, IIb, IIc) and FcγRIII (IIIa, IIIb) have two of these extracellular domains while the high-affinity FcγRIa has three (see Figure 2.37). IgG1 and IgG3 efficiently bind all FcγR types, contributing to their strong effector functions. IgG2, on the other hand, fails to interact with FcγRIa, which may be due to the absence of leucine at position 235 in the low hinge of Fc that is essential for binding to the said receptor. It does bind to FcγRIIa and FcγRIIIa but with low affinity. IgG2 shows enhanced binding to the latter when it is afucosylated. Finally, IgG4 engages FcγRIa at two-fold lower affinity relative to IgG1 and IgG3, and it binds very weakly to the other FcγRs. Thus, it can only activate downstream signaling under multivalent conditions (de Taeye *et al*., 2019).

IgG can also bind type II Fc receptors, such as DC-SIGN and CD23. It also interacts with the neonatal Fc receptor (FcRn) to enable the transcytosis of IgG across the placenta and mediate the long half-life of IgG. TRIM21 is a cytosolic IgG receptor that recognizes IgG-opsonized non-enveloped virus or intracellular bacteria triggering polyubiquitination of the opsonized particle and its subsequent degradation. It also activates the expression of many immunoregulatory genes via the NF-κB, AP-1, and IRF pathways. This TRIM-mediated process is described as antibody-dependent intracellular neutralization (ADIN) that extends the effector functions of immunoglobulins to the intracellular compartments. IgG also engages certain types of Fc receptor-like (FcRL) molecules, which are transmembrane receptors with three to nine extracellular Ig-like domains and several intracellular ITIM and/or ITAM signaling molecules. FcRL5 binds all IgG subclasses, whereas FcRL4 interacts with IgG3 and IgG4 as well as IgA. Both receptors are found on B cells implying that they may regulate humoral immune responses (de Taeye *et al.*, 2019).

IgG is able to assume a hexameric form in complex with C1q so that it initiates complement activation. This multivalent platform resembles the hexameric IgM. It is formed on the cell surface where multiple IgGs are confined into a limited space promoting Fc-Fc interactions between the immunoglobulins. Such interaction is enhanced by the presence of C1q (de Taeye *et al.*, 2019).

Immunoglobulin A

IgA is present in the serum and in secretions. The serum IgA has a molecular weight of 160 kDa and an average serum concentration of 3 mg/ml, making it the second most prevalent circulating antibody after IgG. Meanwhile, the secretory IgA (sIgA) has a molecular weight of 385 kDa and a serum concentration of 0.05 mg/ml. It associates with a glycoprotein, known as the secretory component (SC), that prevents its enzymatic digestion. It is the main antibody in body secretions, like in saliva, tears, colostrum (i.e., 50% of the total protein content), and mucosal secretions in the intestinal, genital, and respiratory tracts (Vaillant *et al.*, 2021; ThermoFisher Scientific, 2021; Schroeder & Cavacini, 2010).

IgA is further classified into IgA1 and IgA2. IgA1 has a longer hinge region with a duplicated stretch of amino acids that is not present in IgA2 (see Figure 3.10). Furthermore, the IgA1 hinge region contains multiple O-linked glycans and two N-linked glycosylation sites per heavy chain. The shorter hinge of IgA2 lacks O-glycans and each heavy chain contains two additional N-glycans. Unlike in the IgG, there are no disulfide bridges in the IgA hinge region, which may provide greater flexibility to IgA. However, the absence of such bonds as well as the elongated hinge region may also increase the sensitivity of IgA1 to bacterial proteases, which may be why IgA2 is present at a greater level in many mucosal secretions (Vaillant *et al*., 2021; Breedveld & van Egmond, 2019; Schroeder & Cavacini, 2010; ThermoFisher Scientific, 2021).

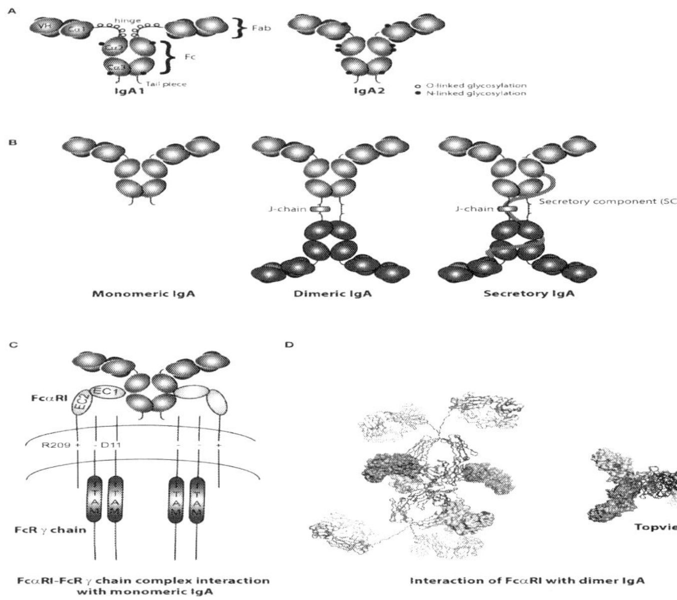

Figure 3.10. (A) Immunoglobulin A (IgA) has two subtypes, IgA1 and IgA2, made up of two heavy chains (blue) and two light chains (brown). IgA1 has a longer hinge with O-linked glycosylation, while IgA2 has more N-linked glycosylation sites. (B) Each subtype may assume monomeric and dimeric conformations. The dimeric form consists of two monomers joined together by a joining chain (J-chain) (green). Secretory IgA is mainly dimeric, and it is associated with the secretory component (red). (Source: Annelot Breedveld & Marjolein van Egmond, Creative Commons License)

These two subtypes may occur as monomers, dimers, tetramers, and polymers (pIgA) (Neurath, 2008). IgA appears mostly as a monomer in the serum, with the monomeric IgA1 present at a much greater concentration in the serum (>90%) than IgA2 (see Figure 3.10). Meanwhile, dimeric form is made up of two monomers that are linked together by the J-chain. This chain is incorporated prior to the secretion of polymeric IgA or IgM. The dimeric IgA1 comprises 40% and IgA2 is 60% in mucosal samples. The secretory IgA(sIgA) is mainly dimeric with some assuming the monomeric or tetrameric structure (Vaillant *et al.*, 2021; de Sousa-Pereira & Woof, 2019).

IgAs, particularly the dimeric forms, confer an essential sort of protection since they are transported to the surface of passages lined with mucous membrane, which include the digestive, respiratory, and genitourinary tracts. These linings have no complement proteins, so the defense mechanism does not involve the complement system. Instead, the IgAs bind the pathogens or their toxins to prevent them from entering the cells of the lining. This gives IgA a critical role in adaptive immunity since the predominant sites for pathogen entry are the mucosal surfaces (Britannica, 2021; Vaillant *et al.*, 2021; de Sousa-Pereira & Woof, 2019).

The release of sIgA into the body secretions is facilitated by pIgR, which transports polymeric immunoglobulins, such as IgM and IgA, across the mucosal epithelium in a process called **transcytosis** (see Figure 3.11). The larger-sized IgM is restricted from diffusing from the serum to the lamina propria so that the smaller IgA is preferentially transferred by pIgR (de Sousa-Pereira & Woof, 2019). Transcytosis is initiated when polymeric IgAs (pIgAs) bind pIgRs that are found on the basolateral surface of epithelial cells lining the mucosal surface. These pIgAs are principally dimers but some are tetramers. The resulting pIgR-pIgA complex is then internalized and transported via vesicular compartments to the apical surface of the cell. The internalized complex can also neutralize pathogens within intracellular vesicular compartments. During transport, the extracellular portion of the pIgR is then cleaved to form the SC that covalently binds, via a disulfide linkage, to the pIgA resulting in the secretory form of IgA (sIgA). SC is a hydrophilic and a highly glycosylated negatively-charged molecule that makes the sIgA the most stable immunoglobulin in secretions since it protects it from proteolytic cleavage after secretion and anchors it to the lining of the mucosal epithelium (Schroeder & Cavacini, 2010; Neurath, 2008; Breedveld & van Egmond, 2019).

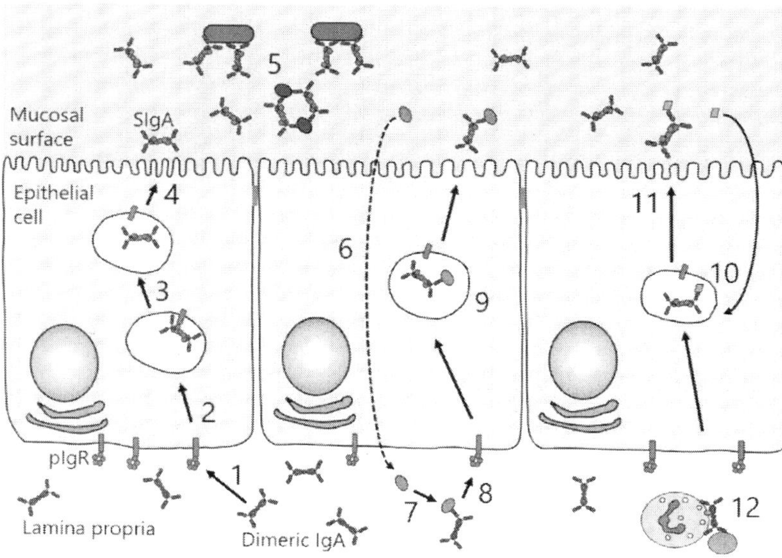

Figure 3.11. IgA is transported across the mucosal epithelium after binding to the polymeric Ig receptor (pIgR). (1) Dimeric IgA (red) produced at the lamina propria of the epithelium binds pIgR (cyan) at the basolateral surface of the epithelial cell. (2) The pIgR-IgA complex is internalized and transported via vesicles across the cell. (3) The secretory component (SC) of the pIgR is cleaved and then associates via a disulfide bond to the dimeric IgA. (4) At the apical surface, the IgA is secreted (i.e., secretory IgA, sIgA) into the lumen. (5) It then binds and neutralizes the pathogens (purple and dark blue). (6) Some pathogens (pink) may penetrate the epithelium and reach the lamina propria. (7) These are bound by dimeric IgA. (8) The pathogen-IgA complex binds to pIgR. (9) After phagocytosis and vesicular transport across the cell, the pathogen is released back into the lumen. (10) Some intracellular pathogens (lime green) may be intercepted by dimeric IgA being transported across the cell. (11) The pathogen is ejected upon release of sIgA at the lumen. (12) Dimeric IgA can also engage phagocytes to engulf and clear pathogens. (Source: Source: Patricia de Sousa-Pereira & Jenny M. Woof, Creative Commons License)

After its release into the mucosal surface, sIgA binds and neutralizes the bacterial, protozoan or viral pathogen present in the luminal space (see Figure 3.11). Specifically, the glycans on IgA interact with sugar-dependent receptors or fimbriae on the surface of pathogens. Since sIgA has multiple antigen binding sites, this leads to agglutination such that the antigen aggregates formed cannot penetrate the mucus lining of the mucosal surfaces. This process is called immune exclusion. The sIgA can also interact with other innate defense factors like mucins, lactoferrin, and the lactoperoxidase system to enhance immune protection. Pathogens that penetrated the epithelium and reached the lamina propria are also bound by dimeric IgA (see Figure 3.11). The resulting pathogen-IgA complex binds pIgR and is then carried across the epithelial cell and then released back into the lumen. Thus, apart from being able to bind IgA alone, pIgR can also interact with IgA complexed with an antigen. This enables the removal of soluble antigens from various origins. It was also observed *in vitro* that some antigens like viral proteins inside the cell, particularly inside endosomes, maybe intercepted by sIgA as it is being transported across the cell. These are released back at the lumen upon the secretion of sIgA so that viral growth is inhibited (de Sousa-Pereira & Woof, 2019).

IgA lacks the binding site for C1q. Therefore, it does not activate the classical pathway of the complement system but activates the alternative pathway possibly via the lectin pathway (see Figure 3.11) (de Sousa-Pereira & Woof, 2019).

Immunoglobulin E

IgE is a monomer with a molecular weight of 188kDa and with the lowest serum concentration of 0.00005 mg/ml. It also has the shortest half-life. Relative to IgG, IgE has additional constant domains in its heavy chain and lacks a hinge region (see Figure 3.12). The Fc of IgE is conformationally flexible assuming an acutely bent form when unbound (Vaillant *et al.*, 2021; Sutton *et al.*, 2019).

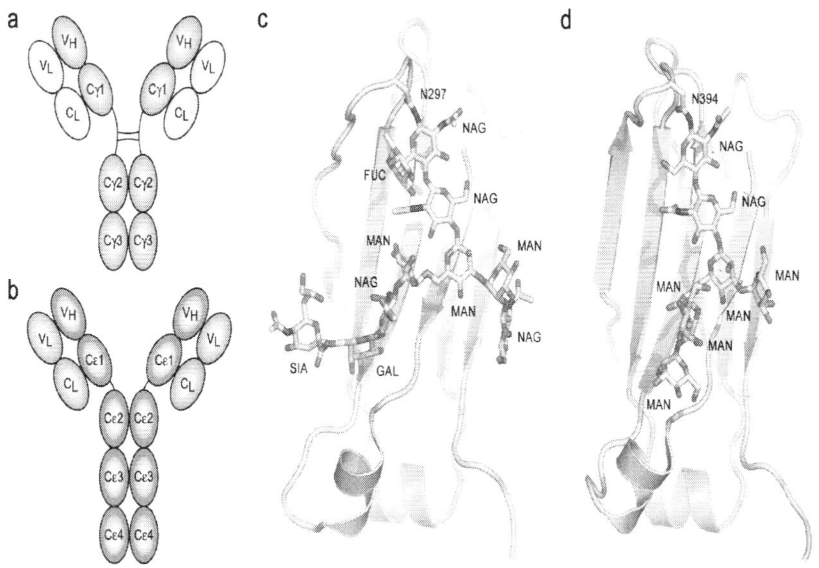

Figure 3.12. The structure of (a) IgG shows some differences compared to that of (b) IgE. The latter has two additional constant domains in each of its heavy chains and it lacks a hinge region. (Source: Brian J. Sutton *et al.*, Creative Commons License)

IgE has two principal receptors, namely FcεRI and FcεRII or CD23. FcεRI is structurally homologous to the FcγR family. It is expressed on mast cells, basophils, airway epithelial and smooth muscle cells, intestinal epithelial cells, APCs, monocytes, and macrophages. IgE has a very high affinity for this receptor so most IgE are already bound to cells. Hence, contact with even a minute amount of allergen can trigger cross-linking of receptor-bound IgEs on mast cells and basophils (see Figure 3.13) inducing cell degranulation. This releases histamine and other vasoactive amines that increase vascular permeability and smooth muscle contraction resulting in an immediate hypersensitivity response that can be powerful enough to cause anaphylactic shock and even death (Sutton *et al.*, 2019).

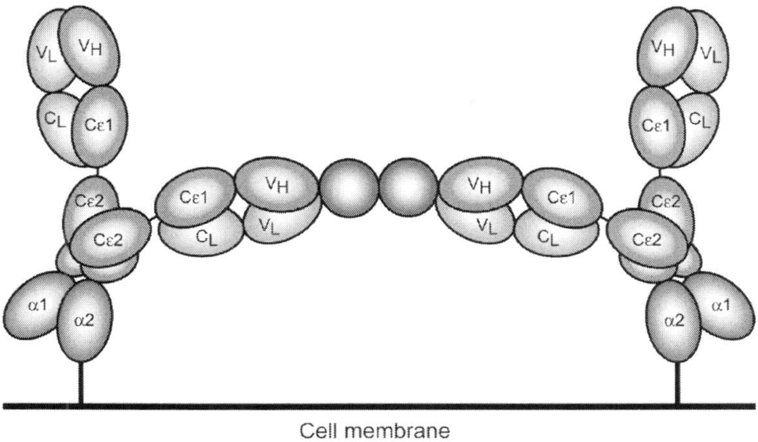

Figure 3.13. An allergen (green) can engage two identical IgE antibodies (yellow and blue) that are linked to their respective FcεRI (purple). This can trigger a cascade of events that result in hypersensitivity and allergic reactions. (Source: Brian J. Sutton *et al.*, Creative Commons License)

FcεRII is a member of the C-type lectin-like superfamily, and it is expressed on B cells, T cells, APCs, gut and airway epithelial cells and other cell types. It is the low-affinity receptor of IgE. Nevertheless, it has three lectin-like heads for IgE so that a high avidity is achieved if more than one head can engage IgE. On B cells, FcεRII can be a membrane protein as well as a soluble protein released from the cell surface, in monomeric or trimeric form, by endogenous or exogenous proteases. Therefore, the binding of IgE to FcεRII is a possible mechanism for IgE homeostasis. FcεRII also transfers IgE-allergen complexes across the gut and airway epithelia promoting the presentation of airborne and food allergens to the immune system (Sutton *et al.*, 2019).

IgE provides defense against parasites, such as *Strongyloides stercoralis*, *Trichinella spiralis*, *Ascaris lumbricoides*, *Necator americanus*, and *Ancylostoma duodenale*. It does so by triggering inflammatory cascades that cause vasodilation and local enhancement of immune responses in cooperation with other immunoglobulin isotypes. Moreover, the binding of IgE to its receptor stimulates eosinophils, platelets, and macrophages to clear parasites via ADCC and ADCP (Vaillant *et al.*, 2021; Sutton *et al.*, 2019).

Immunoglobulin D

Circulating IgD is a monomer with a molecular weight of 184 kDa. Its mean concentration in the serum is 0.03 mg/ml with a short serum half-life, which may be due to the sensitivity of the antibody, especially its hinge regions, to proteolysis. Its structure is similar to that of IgG. It has an unknown function against pathogens since it is not known to be involved in major antibody effector mechanisms. Instead, it is considered to be a BCR. The binding of bacterial proteins to the constant region of IgD triggers lymphocyte activation and differentiation (Vaillant *et al.*, 2021; Schroeder & Cavacini, 2010).

Membrane-bound IgD is expressed on the membrane of B cells that are yet to be stimulated by antigens when these cells mature and leave the bone marrow and populate secondary lymphoid organs. It may have a role in regulating B cell fate at specific development stages through changes in activation status. Membrane-bound IgD is similar to IgM, which functions as BCR and is associated with CD79a and CD79b for signaling. In fact, most IgD^{+} B cells also express IgM and both are involved in BCR signaling. Therefore, in this type of B cells, IgD can replace IgM and *vice versa* (Schroeder & Cavacini, 2010).

Table 3.1. presents the major isotypes of immunoglobulins along with their different properties.

Table 3.1. The Five Major Immunoglobulin Isotypes

Immuno-globulin isotype	Heavy chains	Molecular weight (kilo-Daltons)	Concen-tration in the serum, mg/ml (Percent-age)	Half-life (Days)	Struc-ture	Number of antigen binding sites	Fixes comple-ment	Recep-tor/s
IgM	μ	970->1,000	1.5 mg/ml (5-6 %)	5	Monomer Pentamer Hexamer	2-10	Yes	CR2,CR3 pIgR Fcα/μR, FcμR
IgG	γ	146 (IgG1, IgG2, IgG4) 165 (IgG3)	9 mg/ml (80-85%)	8 (IgG1) 23 (IgG2, IgG3)	Monomer	2	Yes (ex-cept for IgG4)	FcγR DC-SIGN, CD23 FcRn FcRL

IgA	α	160 (serum) 385 (secretory)	3 mg/ml (15%) 0.05 mg/ml (7%)	6	Monomer Dimer Tetramer Polymer	At least 2	Yes	pIgR
IgE	ε	188	0.0005mg/ml (0.02 %)	2.5	Monomer	2	No	FcεRI, FcεRII or CD23
IgD	δ	184	0.3 mg/ml (0.3-1%)	3	Monomer	2	No	CD79a, CD79b

3.2.1.2. Immunoglobulin diversity

The human body can produce more than 10^{12}different types of immunoglobulins even in the absence of antigens. This pre-immune immunoglobulin repertoire is sufficient to ensure that there will be an antigen-binding site to fit any potential antigen although with low affinity. It is after repeated antigen exposure that the B cells produce immunoglobulins that bind their antigens with greater affinity (Alberts *et al.*, 2002). Thus, the repertoire of antibodies present in the immune system at any one time depends on the total number of types of B cells and the antigens encountered by the individual (Janeway *et al.*, 2001; Backhaus, 2018).

Pairing of heavy and light chains

The human genome contains an estimated 30,000 genes, which are significantly lesser than the variety of immunoglobulins that these genes encode. However, the difference among antibodies lies in their antigen-binding sites in the variable region. Therefore, if there are 1,000 genes encoding light chains and 1,000 genes coding for heavy chains, these could theoretically combine to generate 10^6antigen-binding sites (Alberts *et al.*, 2002). The combination of the light and heavy chain genes to generate the variable region was thought to be completely random, but certain studies suggest some preferred pairings between these two chains. However, preferential gene pairing seems to exist only for a small proportion of immunoglobulin loci (Backhaus, 2018).

Somatic germline recombination

Nevertheless, much more variety is produced by the recombination of gene segments making up an immunoglobulin before these are transcribed. This diversification takes place during the earliest stages of B cell development before it leaves the bone marrow and migrates to the secondary lymphatic organs (Aktor, 2012).

Each type of chain in the immunoglobulin, such as κ light chains, λ light chains, and heavy chains, has a distinct pool of gene segments and exons from which a final transcript is formed, and a single polypeptide is produced. Each pool is on a separate chromosome, i.e., IGK locus (κ light chain) is on chromosome 2, IGL locus (λ light chain) is on chromosome 22, and IGH locus (heavy chain pool) is on chromosome 14. Each germline locus contains a large number of gene segments that code for the V region of an immunoglobulin chain and fewer gene segments encoding the C region (see Figure 3.14A). Each V gene segment contains its own promoter, a leader exon, an intron, an exon coding for the first three framework regions (i.e., FR1, FR2, and FR3), CDR1, CDR2, and the CDR3 amino-terminus, and a recombination signal sequence (RSS). Compared to the light chain, the heavy chain locus has several C gene segments corresponding to the different isotypes and subtypes (Alberts *et al.*, 2002; Backhaus, 2018; Schroeder & Cavacini, 2010). These are organized in the order of Cμ, Cδ, Cε, and Cα in the IGH locus (Chi *et al.*, 2020). All C gene segments in the IGH locus can undergo alternative splicing to produce two different types of carboxy-termini, a membrane terminus that attaches the immunoglobulin on the B cell surface or a secreted one that occurs in soluble immunoglobulins. The heavy chain locus also contains the joining (J) gene segment that is made up of its own RSS and an exon encoding the CDR3 carboxy-terminus and FR4. Some gene segments contain frameshift mutations or stop codons so that they do not code for functional products, For example, the IGK locus has 70 V gene segments but only about 30 have been found in functional immunoglobulins (Schroeder & Cavacini, 2010).

While the B cell is still at its progenitor stage in the bone marrow, a complete coding sequence for each of the two chains is assembled by site-specific genetic recombination or somatic recombination of their germline gene sequences. The latter term is used to distinguish it from the meiotic recombination that takes place during gametogenesis and to highlight that it is a DNA rearrangement that occurs in a somatic cell (Alberts et al., 2002; Janeway et al., 2001; Backhaus, 2018). Such DNA rearrangement triggers a shift in the relative positions of the enhancers and silencers to initiate transcription from the promoter. Therefore, a complete chain can be expressed only after the DNA has been rearranged (Alberts *et al.*, 2002). Table 3.2 presents the size and number of functional gene segments of the various immunoglobulin loci.

Table 3.2. The Three Immunoglobulin Loci

Ig chain	Gene locus	Chromosome location	Size (kb)	Number of gene segments			
				Constant (C)	Variable (V)	Diversity (D)	Joining (J)
Heavy chain	IGH	14q32.33	1, 250	9	38-51	27	6
κ light chain	IGK	2p11.2	1, 820	1	34-40	0	5
λ light chain	IGL	22q11.2	1, 050	4-5	29-36	0	4-5

Source: Modified from Backhaus (2018), *Antibody Engineering*, IntechOpen

The joining of the gene segments to form the variable region locus is catalyzed by the enzyme complex V(D)J recombinase made up of two proteins, recombination activating gene(RAG) protein 1 and RAG2, almost exclusively found in developing lymphocytes. Both RAG proteins are large with multiple domains composed of core and non-core regions. RAG1 contains the DNA-binding domain and the catalytic core that mediates the cleavage of the DNA. The function of RAG2 is less understood; it may activate RAG1 to bind and cleave DNA as well as provide additional DNA binding capability. The non-core domains of the RAG proteins generally have regulatory roles (Albert *et al.*, 2002; Chi *et al.*, 2020).

The guided fashion of the V(D)J recombination process is mediated by the recombination signaling sequences (RSS) that serve as recognition sites for the RAG1/RAG2 complex to introduce double-strand breaks (DSBs) (see Figure 3.14C). These are adjacent to the coding region of the gene segments for the variable region (see Figure 3.14A). Each RSS is made up of a conserved heptamer (i.e., seven nucleotides: 5'-CACAGTG-3') that is linked by a non-conserved linker sequence to a conserved nonamer (i.e., nine nucleotides: 5'-ACAAAAACC-3') (see Figure 3.14B). The heptamer is the recognition element with the first three nucleotides (relative to the coding region) showing the highest conservation and essential for recombination. The nonamer sequence is dispensable for recombination with only a few highly conserved nucleotides. The linker sequence is made up of either 12 or 23 nucleotides and it places the heptamer and nonamer sequences on the same side of the DNA molecule separated by either one or two turns of the DNA helix (Backhaus, 2018; Chi *et al.*, 2020; Schroeder & Cavacini, 2010).

A 1-turn RSS with a 12-base pair (bp) linker sequence can only recombine with a 2-turn RSS with a 23-bp linker. This is called the 12/23 rule, which ensures the recombination of the correct gene segments. The

V gene segments of the IGLlocus are always flanked downstream of the coding region(i.e., 3' end)by a 23-bp linker RSS, while the J gene segments are flanked upstream (i.e., 5' end) by a 12-bp linker RSS (see Figure 3.14A). This ensures that only a V gene segment can recombine with a J segment. This order is reversed on the IGK locus, but it still enables V-J joining, i.e., the 12-bp linker RSS is downstream of the V gene segments and the 23-bp linker RSS is upstream of the J gene segments. In the case of the IGH locus, the diversity gene segment is flanked on both sides with the 12-bp linker RSS. In addition, the V and J gene segments are flanked by a 23-bp linker RSS downstream and upstream of the coding region, respectively. The two RSS flanking the D gene allows it to recombine with the J gene and then to the V gene. In contrast, V gene cannot recombine with J gene (Backhaus, 2018; Chi *et al.*, 2020).

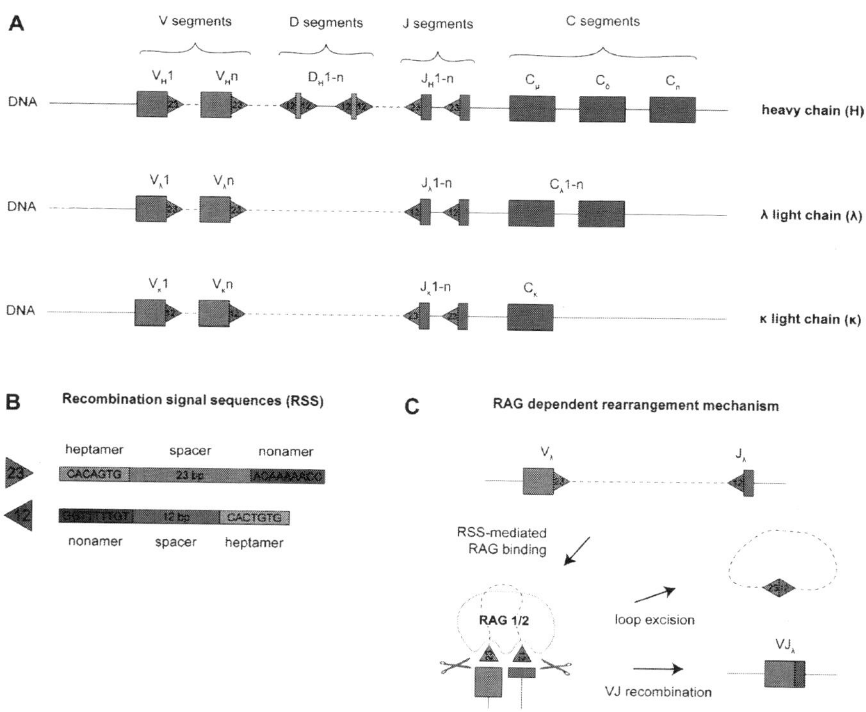

Figure 3.14. (A) The recombination signal sequence (RSS) flanks the gene segments in the locus for the variable region of the immunoglobulin heavy chain (i.e., variable, diversity, and joining genes) as well as the V and J genes in the λ and κ light chain locus. The RSS in these genes varies in terms of orientation as well as the length of their spacer sequence. (B) The RSS has a conserved hep-

tamer sequence and a conserved nonamer sequence with a non-conserved spacer sequence between them. The space can have either 23 base pairs (bp) or 12 bp. (C) During somatic recombination in the variable locus of the λ light chain locus, RAG1/RAG2 complex binds the 23-bp linker RSS adjacent to the V gene and the 12-bp linker RSS adjacent to the J gene. The complex then cleaves the two RSS along with the sequence between V and J (loop excision) resulting in a VJ joined segment. (Source: Oliver Backhaus, Creative Commons License)

In the case of the recombination in the IGL variable region, RAG1/RAG2 complex binds the 23-bp linker RSS adjacent to the V gene and the 12-bp linker RSS adjacent to the J gene (see Figure 3.14C). This brings together the V and J genes in close proximity with the sequence between them forming a hairpin loop possibly by bending the RSS. The RAG1/RAG2 complex then catalyzes the formation of DSBsin both RSS leading to the deletion of the loop so that the V and J genes are fused together via non-homologous end joining mechanism(NHEJ) (Backhaus, 2018). The intervening DNA between the two DSBs may not be deleted but inverted instead. Whether the recombination proceeds in a deletional or inversional manner depends on the relative orientation of the two RSS. If the V and J segments are in the same nucleotide strand, these are recombined by deletion, while if these are on opposite strands, they are joined by an inversion between the RSS (Chi *et al*., 2020).

In the final transcript produced after recombination, the C region is encoded by a single segment of DNA, whereas the V region is a continuous exon with two or more segments. In particular, two gene segments form the DNA sequence for the V region of a light chain. These are the long V gene segment and a short J segment. In the unrearranged DNA sequence for a light chain, the C region is located quite a distance from the V gene segments, whereas the J gene segments are close to the C region. The joining of a V gene segment to a J gene segment during recombination brings the V gene closer to the C region. After recombination, the J gene segment is separated from the C region only by one intron. For the IGH V region, it is encoded by a DNA sequence assembled from three gene segments, namely V, J, and diversity or D segments, with the D gene segment located between the V and J segments (see Figure 3.15). This is why the recombination process involved during B cell development is also called V(D)J joining. The recombination process to produce a complete heavy chain C region has two stages. First, a D gene segment is joined to a J segment. Then a V gene segment recombines with the DJ segment (Janeway *et al*., 2001; Alberts *et al*., 2002). The IGH locus is then transcribed with the Cμ and the

Cδ gene segments. The primary mRNA then undergoes alternative splicing so that when the Cμ exon remains in the transcript then the final mRNA is translated to IgM heavy chain. On the other hand, if the Cδ is joined to the variable region exon, then the mRNA codes for IgD heavy chain (Backhaus, 2018).

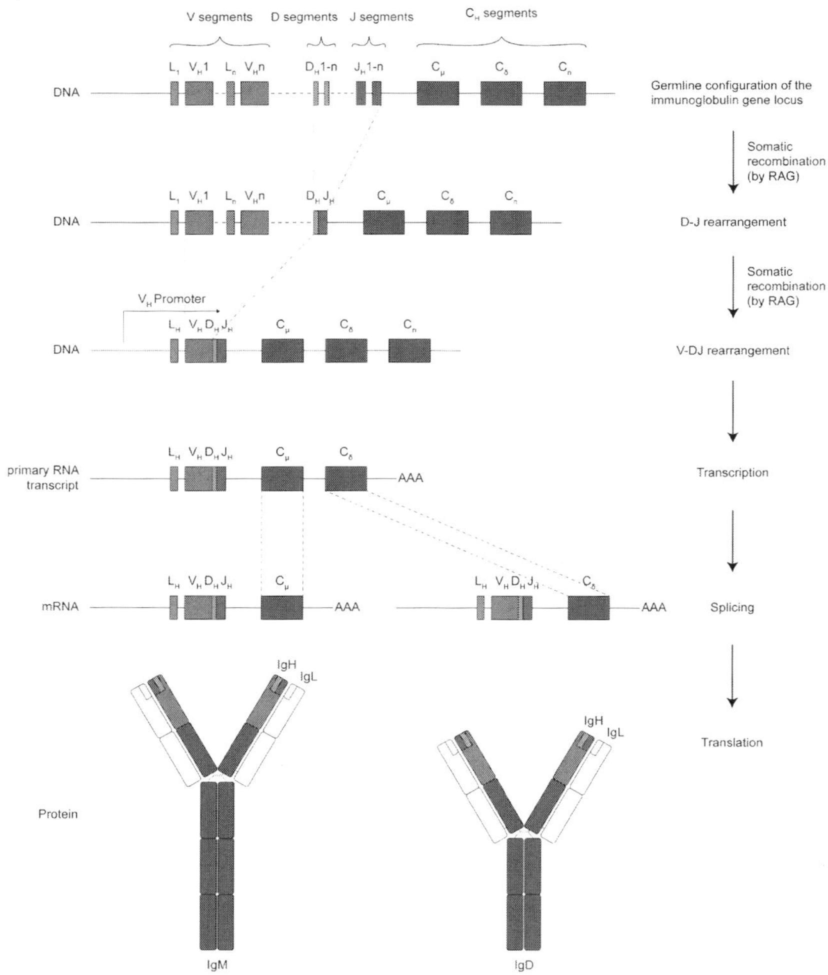

Figure 3.15. The germline IGH (immunoglobulin heavy chain) locus contains several gene segments for the constant (C) and variable (V) regions. Somatic recombination by RAG proteins joins the diversity (J) and joining (J) segments first then the variable (V) segment is joined next. After transcription, C segments in the primary transcript are spliced and one of them is recombined with the VDJ segment to form the final mRNA. (Source: Oliver Backhaus, Creative Commons License)

There are a large number of V, J, and D gene segments that can be combined in various ways to produce very diverse immunoglobulins. For example, any of the 40 V segments in the κ light chain gene pool can be randomly joined by the RAG1/RAG2 complex to any of the five J segments to produce 200 possible V regions in the κ light chain. In the case of the λ light chain, it can have at least 116 different V regions. There are 51 V segments in the heavy chain gene pool that can combine with any of the six J segments and any of the 27 D segments. This can generate at least 8,262 different heavy chain V regions. The 316 different V regions of the light chain can combine with the 8,262 different heavy chain V regions to generate about 2.6×10^{6}different antigen-binding sites. Then the heavy chain and light chain assembly significantly increases the number of possible combinations by more than 10^{8} fold to account for the huge number of antibody types in the body (Alberts *et al*., 2002).

Junctional diversification

The site-specific recombination of the immunoglobulin gene segments via NHEJis error-prone. A number of nucleotides at the ends of the combined gene segments (i.e., at the VD and DJ junctions) are usually lost and are replaced by one or more random nucleotides (Blair & Bosma, 2021). The process of losing and gaining these randomly chosen nucleotides is called junctional diversification, and it greatly increases the diversity of the Vregion coding sequences that are produced by recombination. Specifically, the CDR3 diversity is greatly influenced by this process because of its position between the V and J gene segments (Blackhaus, 2018). Junctional diversity is the major source of variation in the pre-immune repertoire (Schroeder & Cavacini, 2010). However, this also causes a mutation that may produce a non-functional gene so that the B cell fails to produce a functional immunoglobulin. Such B cells eventually die in the bone marrow (Alberts *et al*., 2002).

When the RAG1/RAG2 complex excises the intervening DNA and mediates recombination between two gene segments, it leads to loss of one to several nucleotides and produces short hairpins on the blunt ends of the gene segments (see Figure 3.16). Then the Artemis or DNA-dependent protein kinase (DNA-PK) complex is recruited and introduces a single-stranded break (SSB) at random sites in the hairpins. In many cases, this can produce a 3' overhang of palindromic DNA sequences, which are called P nucleotides. Then the terminal deoxynucleotidyl-transferase (TdT) adds up to 20 nucleotides to the single-stranded P nucleotide stretch. The additional nucleotides are called N nucleotides (i.e., non-template) because these were randomly added without a template. Some nucleotides in the two single-stranded stretches are complementary so that they can hybridize via the formation of hydrogen bonds between base pairs. The mismatched nucleotides are removed by an exonuclease and the remaining gaps are filled by a DNA polymerase with bases complementary to the opposite strand. Finally, the nicks in the DNA strand backbone are joined together by the DNA ligase IV/X-ray repair cross-complementing protein 4 (XRCC4) complex (Backhaus, 2018).

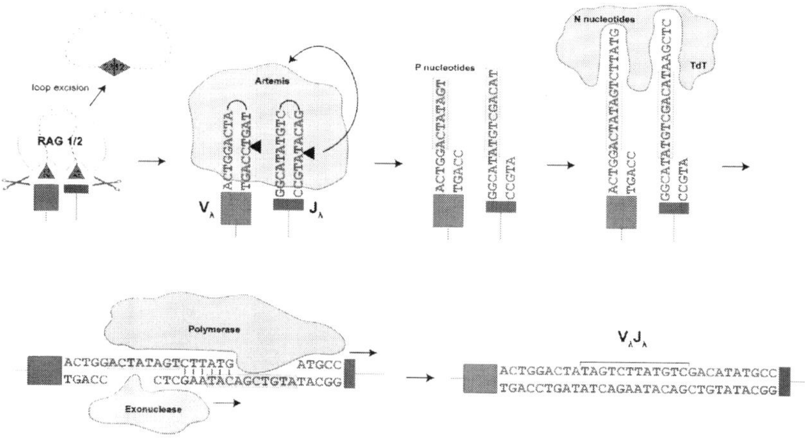

Figure 3.16. After RAG1/RAG2 removes the intervening sequence between the variable (V) and joining (J) genes of the λ light chain, hairpins are formed at the excision site. The Artemis complex then makes a single-stranded cut in both hairpins producing palindromic DNA sequences. These so-called P nucleotides are extended by the addition of random nucleotides, called N nucleotides, by the terminal deoxynucleotidyl-transferase (TdT). The complementary bases in the two single-strand segments pair together, while the mismatched nucleotides are removed by an exonuclease. The gaps are then filled by a DNA polymerase and the breaks in the DNA strands are ligated. (Source: Oliver Backhaus, Creative Commons License)

The imprecise joining process and variation in the number of N nucleotides added can produce up to 10^7 heavy chain CDRs with different lengths and structures (Schroeder & Cavacini, 2010). The addition of N nucleotides differs in frequency between the heavy chain and the light chain locus, with the former showing higher frequency. This is due to the higher expression level of the TdT when the heavy chain locus is rearranged than when the light chain locus undergoes recombination (Backhaus, 2018). The addition of nucleotides can cause frameshift mutations in the coding sequence beyond the junction. The disruption of the reading frame often results in the production of nonfunctional immunoglobulins. Therefore, the achievement of junctional diversity comes at a high cost (Janeway *et al.*, 2001).

Allelic exclusion

Humans have two copies of each gene including the final version of immunoglobulin heavy chain and light chain mRNAs generated after recombination and junctional diversification. Both copies of the final mRNAs are transcribed, but usually, only one of them is translated and assembled into a functional BCR. This is called **allelic exclusion**, wherein only one of the two alleles is expressed on the B-cell surface (Backhaus, 2018). Therefore, each B cell expresses only one type of heavy chain and light chain. Then when this B cell divides, all of its progeny cells have the same antigenic specificity (Aktor, 2012).

One possible mechanism for allelic exclusion that is proposed is that during somatic germline recombination, only one locus is chosen for recombination, while the other is silenced. When a functional heavy chain is already generated, the expression of RAG1/RAG2 is downregulated and the complex is targeted for degradation. Also, the access of RAG1/RAG2 to the heavy chain loci is restricted and such restriction is sustained even when the light chain locus is undergoing rearrangement. This ensures that no further rearrangement or change of allele activity will occur (Backhaus, 2018).

However, when the expressed allele did not produce a functional BCR, the second one is activated and expressed. Having a choice of two different immunoglobulin alleles further increases diversity. When the second allele also produces a non-functional or less efficient BCR, the B cell dies by apoptosis. This process is called **clonal deletion** (Backhaus, 2018).

B cell receptor editing

In order to escape clonal deletion, the allele with a non-functional product may also undergo additional rounds of V(D)J recombination until a functional BCR is produced. This may also occur when a BCR may be functional but against self-antigens. This process is called receptor editing, which serves as a checkpoint and rescue mechanism to correct the non-functional or self-reactive BCR. This involves rearrangements of the V and J gene segments in the light chain or the replacement of a V gene segment in the heavy chain (Backhaus, 2018).

Somatic hypermutation

After the synthesis and assembly of the heavy and light chains of immunoglobulins acting as a functional BCR on the cell surface, the naïve B cells migrate to the secondary lymphatic organs. At this point, the BCR can recognize a diverse range of antigens but with low affinity. Still, such low affinity interactions are enough to induce a primary immune response. Then, the B cells undergo rapid cell divisions in the germinal centers of the lymph node, during which the affinity of the BCRis progressively enhanced in a process called affinity maturation. This is the result of the accumulation of point mutations in the V region coding sequence of both the light and heavy chains. This takes place when the B cells are stimulated by antigen and Th cells over a period of few months resulting in affinities that are greater by 100-fold. The mutation rate involved is about one per Vregion coding sequence per cell generation. This is a million times higher than the spontaneous mutation rate in other genes (i.e., 10^{10}mutations per cell cycle) that is why this process is called somatic hypermutation (Blair & Bosma, 2021; Alberts *et al*., 2002; Backhaus, 2018; Reverberi & Reverberi, 2007).

These random mutations are catalyzed by activation-induced cytidine deaminase (AID), which is a 198- amino acid protein that catalyzes the deamination of cytosine in single-stranded DNA to uracil. However, AID can only deaminate 3% of the cytidines even at hotspots (Chi *et al*., 2020). The conversion of cytosine to uracil leads to mismatch pairing between uridine and guanosine. This is detected by base excision repair (BER) and mismatch repair (MMR) proteins. In the case of the mismatch repair pathway, mutSα is made up of mutS homolog 2 (MSH2) and MSH6 detects the mismatch. This leads to the recruitment of apurinic/apyrimidinic endonuclease 2 (APE2) and exonuclease 1 (Exo1). APE2 creates a nick 5' of the mismatch, which serves as the entry point for the Exo1 to excise the uridine and the adjacent nucleotides. Proliferating cell nuclear antigen (PCNA) subsequently recruits DNA polymerase η (DNA polη) to fill in the gap. This enzyme is error prone in B cells but has no exonuclease activity. It incorporates thymidine regardless of the template sequence leading to a preference of adenosine-thymidine mutations at the original deaminated site and the adjacent nucleotides. Ligase 1 (Lig1) finally seals the new strand with the existing strand to finish the repair. In the BER mechanism, the deaminated cytosine is detected by the uracil DNA glycosylase (UNG or UDG) cleaving the uracil from the uridine to leave an abasic site in the DNA strand. This site now serves as a non-informative template during DNA replication. PCNA recruits the DNA polymerase REV which inserts a random nucleotide in the growing DNA strand opposite the abasic nucleotide. After another round of DNA replication, there is a stable transverse mutation at the site of the original C: G pair (Blackhaus, 2018; Chi *et al.*, 2020).

When the B cells migrate to the lymph nodes, their expression pattern for certain genes changes depending on whether they are in the light or dark zone of the germinal center. In the dark zone, the B cells greatly express CXC chemokine receptor type 4 (CXCR4) (see Figure 3.17). The expression is reduced when the B cells are in the light zone. The ligand of CXCR4 is CXCL12, which is expressed on the surface of reticular cells in the dark zone. The CXCL12/CXCR4 signaling of B cells is considered as a homing signal to keep B cells in the dark zone if CXCR4 expression is high. The dark zone is where affinity maturation of BCR occurs (Blackhaus 2018).

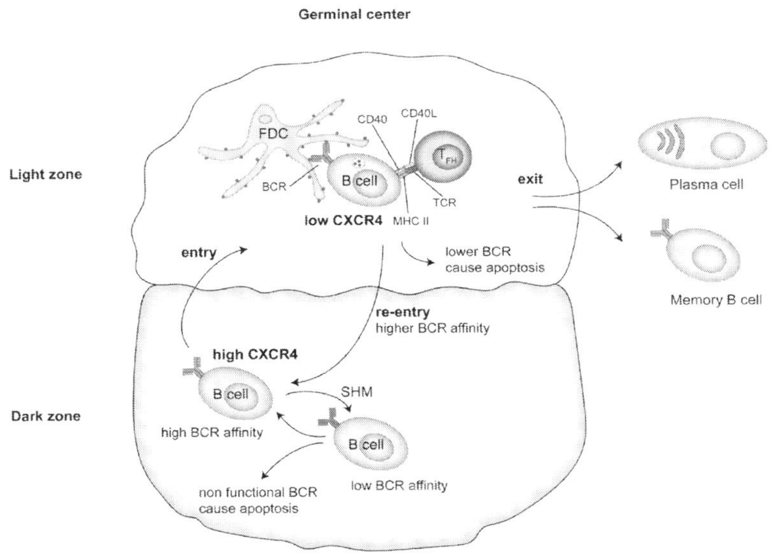

Figure 3.17. B cells in the dark zone of the lymph node germinal center highly express the chemokine receptor CXCR4 and undergo somatic hypermutation in the variable region genes of their immunoglobulin loci. When they enter the light zone, CXCR4 expression is downregulated. Follicular dendritic cells (FDCs) in the light zone present foreign antigens, which are bound by specific B cells with B-cell receptors (BCRs) that have affinity to them. The resulting antigen/BCR complex is processed, and the generated antigen peptides are presented by major histocompatibility complex II (MHC II) to the T cell receptor

(TCR) of follicular helper T (T_{FH})cells. The CD40-CD40L interaction and cytokines from T cells provide survival and proliferation signals to the B cells. On the other hand, the low-affinity B cells undergo apoptosis. The surviving B cells then reenter the dark zone where they can undergo another cycle of somatic hypermutation. This cyclic affinity maturation process produces B cells with high-affinity BCRs. These B cells finally leave the light zone and differentiate into plasma cells or memory cells. (Source: Oliver Backhaus, Creative Commons License)

The point mutations can change one or a few amino acids in the CDRs or the V region. This may dramatically change the antigen specificity and affinity of the resulting immunoglobulin. Only a small proportion of BCRsthat are produced by hypermutation manifest increased affinity for the antigen that is presented by follicular dendritic cells (FDCs) in the light zone (see Figure 3.17). The B cells expressing these receptors can then capture and process more of the antigens and present the antigen peptides via the MHC II to the follicular helper T cells (Tfh). Therefore, these B cells receive more survival and mitogenic signals through TCR recognition, CD40-CD40L interaction, and cytokine stimulation from T cells. As a result, these B cells preferentially survive and proliferate or are positively selected to mature, while B cells with low-affinity or non-functional BCRs die by apoptosis or are negatively selected (Backhaus, 2018).

The positively selected B cells reenter the dark zone and upregulate CXCR4 expression. Another round of somatic hypermutation may occur, which may further increase BCR affinity. The B cells may undergo affinity maturation repeatedly until they express high-affinity BCRs. They finally leave the light zone and then differentiate either into plasma cells that secrete antibodies with the same specificity as the BCR or into memory cells. Therefore, repeated cycles of somatic hypermutation and antigen-driven proliferation of certain memory B cells result in high-affinity immunoglobulins becoming more abundant during an immune response. For this reason, affinity maturation is also called the cyclic reentry model (Alberts *et al*., 2002; Backhaus, 2018).

3.2.1.3 Class-switch recombination

In the bone marrow, all B cells initially synthesize IgM. Then after they leave the bone marrow but before they encounter any antigen, they produce both IgM and IgD as BCRs with the same antigen-binding sites. Upon stimulation by antigens and activation by Th cells, B cells primarily produce

IgM so that this isotype dominates the primary immune response (Alberts *et al.*, 2002). Later in the immune response, antigens and cytokines trigger a process called class-switch recombination (CSR) or isotype switching, wherein immunoglobulin isotypes switch from IgM or IgD to IgG, IgA or IgE. This is achieved by exchanging the constant region in the heavy chain locus, while the variable region remains unchanged. The switch enables the immunoglobulin to interact with different effector molecules via their Fc region (Backhaus, 2018).

CSR is a chromosomal DNA rearrangement that is guided by a conserved **switch(S)regions**, which are repetitive DNA segments upstream of the C gene segments of the heavy chain except for Cδ (see Figure 3.18). Apart from the S region, each C gene segment is also preceded by an I exon, which cannot be translated. Certain cytokine or extracellular signals can induce its transcription, which activates its associated gene segment. The transcription of a V gene or a S region upstream of a C gene opens up the double-stranded DNA producing single-stranded DNA that can now be accessed by AID, which introduces SSBs on two S regions (Schroeder & Cavacini, 2010). It is also thought that if two uridines are close to each other in the two opposite strands, a DSB may result through the BER pathway. Conversion of two more distal uridines into a blunt DSB requires MMR. The DNA between the two breaks is excised and this excised region always includes the μ and the δ constant gene segments. The two DNA fragments are then brought together via NHEJ mechanism. This combines the V gene segments with the C gene segments that remained (Backhaus, 2018; Chi *et al.*, 2020).

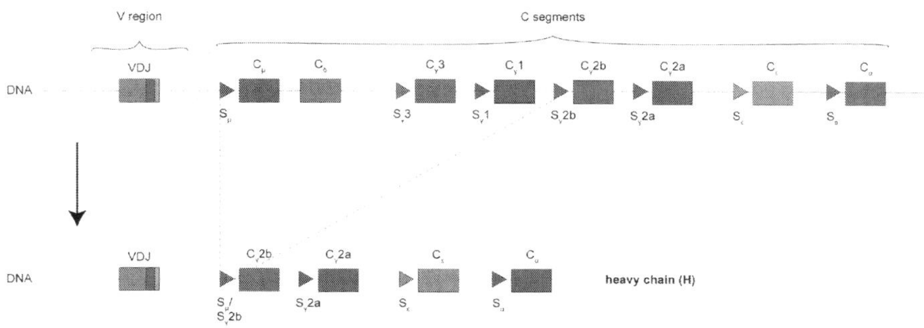

Figure 3.18. Class-switch recombination is induced by activation-induced cytidine deaminase (AID) producing a single-stranded break on two switch (S) regions upstream of constant gene segments of the heavy chain locus. The DNA between the cuts is removed and non-homologous end joining machinery (NHEJ) joins the remaining constant gene segments to the V(D)J segment. (Source: Oliver Backhaus, Creative Commons License)

Certain conditions can also provide downstream signals for CSR. For example, IgG1 and IgG3 are mainly produced in response to protein antigens, but chronic stimulation with these antigens favors the production of IgG4. The isotype produced by CSR is also influenced by different cytokines secreted by T cells since different promoters located upstream of each acceptor S region are activated by different cytokines IgE and IgG secretion is induced by IL4, while IgG1 or IgG3 synthesis is activated by IL10. Moreover, TGFβ initiates the conversion to IgA, and IFNγ induces IgG2 or IgG3 production (Backhaus, 2018; Chi *et al.*, 2020).

IgE, in particular, has several inducing factors so that it has a higher probability to be transcribed relative to IgG1. Allergen stimulation can damage epithelial cells so that they produce thymic stromal lymphoprotein. This protein activates DCs triggering naïve CD4 T cells to differentiate into inflammatory Th2 effector cells that secrete IL4, IL9, and IL13 that enhance IgE synthesis. DCs also activate allergen-specific Th2 memory cells, which mediate CSR from IgM to IgE. Also, microRNA-146a upregulates the expression of 14-3-3σ, which is an important CSR protein, enhancing CSR into IgE (Chi *et al.*, 2020).

3.2.2 Recognition of Antigens by Immunoglobulins

When a pathogen enters the body, its component antigens bind with B lymphocytes whose BCRs are complementary to their shape. Immunoglobulins bind antigens with specificity and high affinity via the Fv region in the Fab. In particular, it is the paratope, the specific part on the Fv region of the immunoglobulin where the antigen binds, that interacts with the epitope, the specific site or the distinct three-dimensional pattern on the antigen that is bound (see Figure 3.19). Antigens may be viewed as a collection of epitopes as they can have many epitopes on different parts of their three-dimensional protein structure. Therefore, an antigen may be bound by different immunoglobulins via its various epitopes (see Figure 3.20). Immunoglobulins usually recognize surface epitopes independent of the other components of intact antigens. This enables immunoglobulins to differentiate closely related antigens from each other. On the other hand, it also allows one immunoglobulin to bind different antigens that have similar epitopes. This phenomenon is called cross-reactivity (Schroeder & Cavacini, 2010). This enables an immunoglobulin to bind an antigen that is different from the one that elicited its synthesis and secretion. Recognition of the appropriate epitope by the BCR stimulates the B cell to divide and generates selected clones whose BCRs have identical specificities. These clones further differentiate into plasma cells that secrete a specific immunoglobulin (Neurath, 2008). Such immunoglobulin is identical or closely similar to the BCR of the B cell that originally recognized the antigen. In any case, the immunoglobulins formed can recognize and bind the invading antigen.

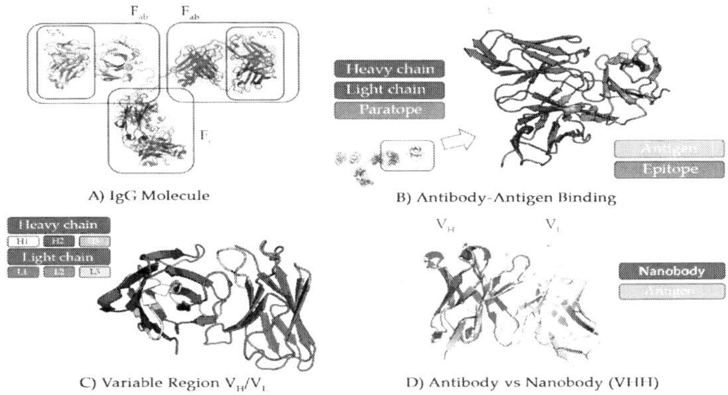

Figure 3.19. The immunoglobulin paratope (light green) binds to the antigen epitope (orange). (Source: Richard A. Norman *et al.*, Creative Commons License)

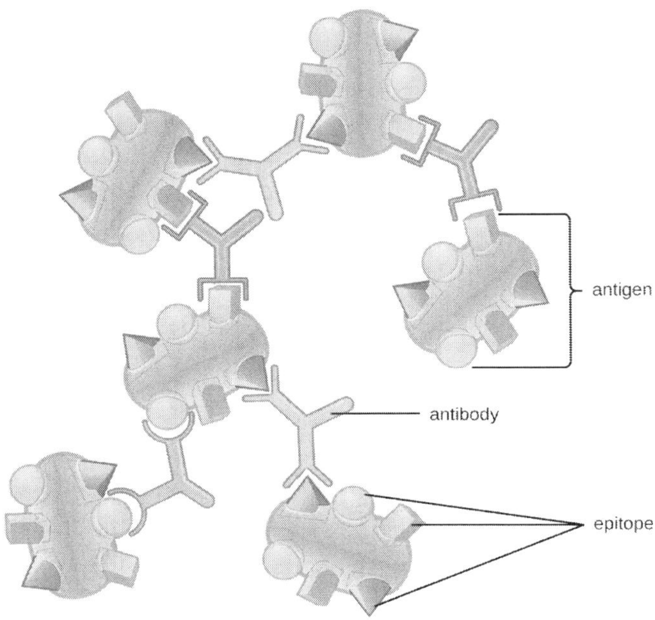

Figure 3.20. A proteinaceous antigen is typically composed of multiple epitopes so that it can be recognized by different immunoglobulins or antibodies. (Source: OpenStax, Creative Commons License)

3.2.2.1. *Paratope structure*

The paratope is made from the light and heavy chain variable domains folded in a way that brings the six hypervariable loops (i.e., three in the light chain and three in the heavy chain) together (see Figure 3.21). The immunoglobulin Fc region, on the other hand, is composed of the heavy chain constant domains CH2 and CH3, and it mediates the biological activity of the immunoglobulin (Sela-Culang *et al.*, 2013).

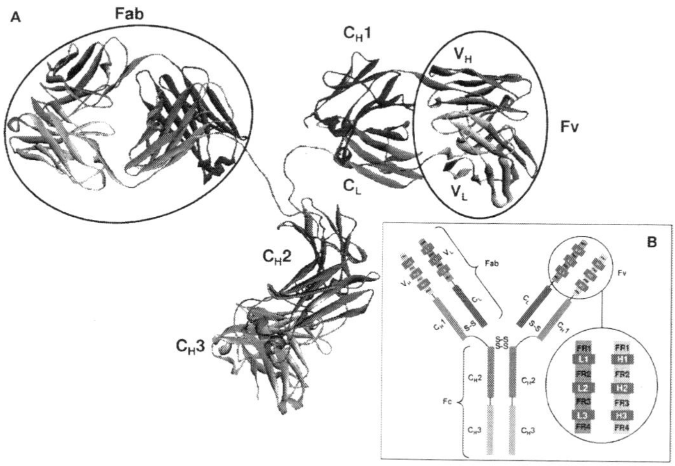

Figure 3.21. (A) The Y-shaped arm of the immunoglobulin is known as the antigen binding fragment (Fab), which is composed of the two variable domains of the heavy (VH) and light chains (VL) as well as the two constant domains (CH1 and CL). VH and VL form the Fv fragment. (B) This fragment contains the paratope consisting of the three hypervariable loops in the VL (i.e., L1, L2, and L3) and three in the VH (i.e., H1, H2, and H3). (Source: Inbal Sela-Culang *et al.*, Creative Commons License)

The identification of paratopes is usually done by the identification of three regions in the hypervariable regions known as CDRs. This is based on the assumption that CDRs are the most variable regions among immunoglobulins. In fact, the amino acid residues that directly interact with the antigen are generally the most variable ones and are located in the CDRs. The CDR loops are flexible and can adopt various distinct conformations, which influence the binding properties. Among the three CDRs, the hypervariability, in terms of sequence and length, of the CDR3 is critical in determining the specificity of immunoglobulins. As discussed previously, such variability is generated by somatic recombination of the V, D, and J gene segments as well as by somatic hypermutation. Certain amino acids are preferentially enriched and depleted in the antigen binding regions. Nevertheless, not all amino acid residues within the CDR bind the antigen. These non-contacting residues are important for maintaining the structural conformations of the hypervariable loops (Sela-Culang *et al.*, 2013; Akbar *et al.*, 2021).

Unlike paratopes, epitopes have an amino acid composition that is similar to that of non-epitopic surface residues. However, if the epitope residues are divided into six subsets based on the CDR they preferentially bind, the subsets have a distinct amino acid composition that is different from the non-epitope surface (Sela-Culang *et al.*, 2013).

There are also non-CDR regions known as the framework regions (FRs) that are scaffolds for the CDRs, but they also contribute to antigen recognition (see Figure 3.21B). FR amino acid residues that influence antigen binding have two categories. The first are those that contact the antigen as part of the binding site. These are closely similar in amino acid sequence to the CDRs and may even be within the boundaries of the CDRs. Some of the amino acid residues belonging to this category significantly differ in sequence from the CDRs but are near to them in the three-dimensional form. The second category is made of FR residues that are not in contact with the antigen but can still indirectly affect antigen binding. Some of these are located close to the CDRs and affect binding by providing a structural support to the CDRs so that they can adopt the right conformation and orientation required for antigen binding. Those that are far from the CDRs may still maintain the overall form of the Fv domains. They may also influence antigen binding by directing the orientation of VH relative to VL, and thus the orientation of the CDRs relative to each other (Sela-Culang *et al.*, 2013).

The constant domains of immunoglobulins are well-known for being responsible for differences in isotype and for the immunoglobulin's effector functions, such as complement activation, Fc receptor binding, avidity, and serum half-life. It turns out that they may also affect antigen binding. This is proven by the fact that there are many immunoglobulins with identical variable domains, but they bind the same antigen with different affinity or specificity because they belong to different isotypes. This may be because the constant domains exert an allosteric influence on the structure of the variable domains. Changes in the constant domain amino acid sequence may result in the rearrangement of the constant domains relative to each other and relative to the variable domains. This may modify the orientation of VH and VL relative to each other, thus re-shaping the antigen binding site (Sela-Culang *et al.*, 2013).

3.2.2.2. Paratope-epitope interactions

The binding of a paratope with its epitope is mediated by weak interactions making it a reversible process. For the binding to be stable, several weak bonds must be present and there must be steric complementarity between the molecules in order to overcome the overall repulsion between the two particles. The actual binding site involves a few amino acids (or monosaccharides in certain antigens) in very small regions of the molecules. When the paratope and the epitope are several nanometers apart, they are attracted together by long-range forces like ionic and hydrophobic bonds. These expel water molecules between the two molecules so that they approach each other more closely. The van der Waals forces can then form between them, but ionic groups still play a role (Reverberi & Reverberi, 2007). Paratopes may detect epitopes in two ways (see Figure 3.22). They may interact with conformational epitopes, which retain their native state on a fully folded antigen. On the other hand, paratopes may bind linear epitopic elements, which are continuous linear segments of antigenic amino acid residues. A linear epitopic element may correspond to the full epitope sequence or only a part of it. Still, it folds into a structure resembling a conformational epitope when it is bound to a paratope (Ledsgaard *et al.*, 2018).

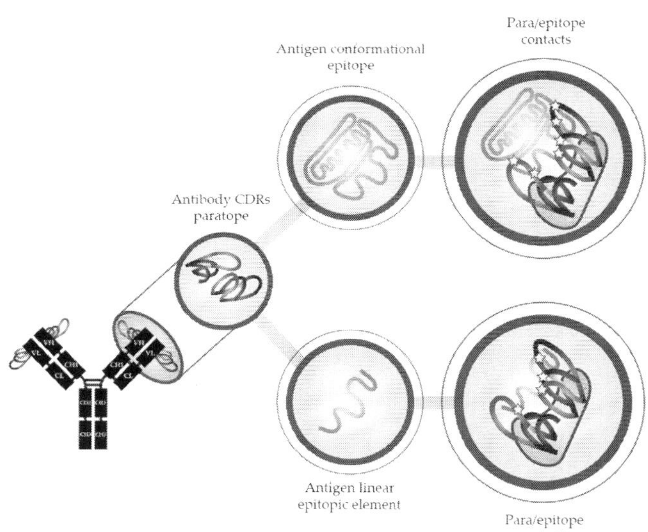

Figure 3.22. The paratope of an immunoglobulin may bind the entire antigen via the conformational epitope. Alternatively, a linear ectopic element of the said epitope may be bound in a similar manner but with fewer points of interac-

tion (stars). (Source: Line Ledsgaard *et al.*, Creative Commons License)

The paratope of early immunoglobulins is cross-reactive or binds with different antigens via low affinity interactions. As the immunoglobulins undergo affinity maturation, there is a decrease in the conformational flexibility of their antigen binding sites leading to greatly enhanced paratope-epitope binding (Reverberi & Reverberi, 2007).

The interaction between paratope and epitope may be affected by several factors, including temperature and pH. The specific effect of temperature depends on the nature of the epitope and paratope as well as the type of weak bonds involved. For instance, hydrogen bonds, which are essential when the antigen is a carbohydrate, are more stable at low temperatures. On the other hand, the strength of hydrophobic bonds increases with temperature (Reverberi & Reverberi, 2007).

A change in a few amino acids, a process called antigenic drift, may produce a new epitope. Correspondingly, immunoglobulins also undergo adaptive mutations in the paratope area via modifications of their amino acid sequence or structural variations (Qiu *et al.*, 2015).

While the binding of the antigen by the immunoglobulin is a critical process in adaptive immune responses, it does not necessarily translate to neutralization of the bound antigen. Many studies reported a positive correlation between affinity and neutralization, but others demonstrated that this is not true for all antigens. This lack of correlation may be due to differences in neutralization efficiencies of paratopes on different epitopes (Ledsgaard *et al.*, 2018).

3.2.3 B Lymphocyte Activation

B cell development continues to occur upon the influence of antigens. The specific antigen and cytokines will determine whether the activated B cell will either become a plasma cell that produces large amounts of immunoglobulins or a memory cell that can be activated again in the future (Bonilla & Oettgen, 2010).

3.2.3.1 T-independent antigens

T-independent (TI) antigens can induce B cell proliferation and immunoglobulin formation without the assistance of T cells. In mice, these can be further classified as TI type 1 antigens that can induce on their own the proliferation and immunoglobulin production of mature B cells. An example is plant lectin, such as pokeweed mitogen, lipopolysaccharide, or endotoxin. At lower doses, TI type 1 antigens stimulate the division of B cells that secrete antigen-specific immunoglobulins. Relatively high doses of these antigens activate TLR4 and its MyD88-mediated signaling pathway resulting in the polyclonal proliferation of B cells. Another subset is the TI type 2 antigens. These are macromolecules, whether proteins, nucleic acids, glycolipids, or polysaccharides, with repeating molecular patterns that can bind several BCRs and cross-link them. Examples of this antigen type include the capsular polysaccharides in bacteria as well as their flagella (Britannica, 2021). They behave like TI type 1 antigens at low doses. These could provide a partially activating signal to B cells in addition to the signals from cytokines or from contact with other cells like DCs (Bonilla & Oettgen, 2010). In particular, the BCR cluster, consisting of 10-20 receptors, induces local membrane association of multiple activated Bruton's tyrosine kinases (Btk) molecules resulting to long-term mobilization of intracellular ionized calcium. The persistent movement of calcium recruits transcription factors, thereby inducing B cell activation and proliferation. A second signal is required for the secretion of immunoglobulins. This could be provided by the activation of TLRs when they recognize the antigen. The second signal can directly target B cells or induce other immune cells to produce cytokines or co-stimulatory molecules (Vos *et al*., 2000). One possible second signal may be provided by a complement molecule. TI type 2 antigens can bind the complement fragment C3d, which in turn is bound to CR2 (or CD21). This results in cross-linking of BCR and CD21 on polysaccharide specific B lymphocytes (Jeurissen *et al*., 2004).

TI antigens lead to rapid production of IgM and IgA with low antigen-binding capacity. They do not trigger isotype switching, germinal center formation,affinity maturation as well as immunological memory formation (Neurath, 2008; Rodgers & Rich, 2008). However, some studies show that TI antigens induce CSR in mice, but lower concentrations of IgG are formed. Also, germinal centers have been formed in response to certain TI antigens, but these were short-lived with little evidence of somatic hypermutation (Allman & Northrup, 2010).

This mode of B cell activation involves co-stimulatory molecules or ligands such as a cytokine known as B-cell activating factor (BAFF) (also called B lymphocyte stimulator/BLyS, TALL-1, THANK, zTNF4)and its homologue, a proliferation-inducing ligand (APRIL). BAFF is expressed on DCs and myeloid cells and plays an important role in B cell differentiation and proliferation. There are three known BAFF receptors expressed on B cells, namely BAFF-R, transmembrane activator, calcium modulator, and cyclophilin ligand interactor (TACI), and B cell maturation factor (BCMA). APRIL is expressed on a wide range of leukocytes including activated T cells, and it is recognized by TACI and BCMA. Apart from co-stimulating B cell activation, it also enhances the antigen presentation ability of B cells, increases B cell survival period, and regulates tolerance. The direct interaction between DCs and B cells activates the signaling pathway mediated by these co-stimulatory molecules. Along with the signals from TI antigens, this signaling system can trigger CSR without the involvement of T cells. This process provides a rapid and adequate immune response to polysaccharide antigens prior to the recruitment of T cell-mediated response (Jeurissen *et* al., 2004; Bonilla& Oettgen, 2010; Cano & Lopera, 2013).

3.2.3.2 T-dependent antigens

Most of the responses to antigenic proteins and glycoproteins mediated by immunoglobulins involve the cooperation of T cells. These T-dependent antigens may contain epitopes that T cells recognize and others that B cells identify. These epitopes are presented to the Th cells by APCs or the B cell itself. T-dependent antigens with low molecular weight may diffuse directly into B cell areas in secondary lymphoid tissues, such as the primary follicles in the lymph node cortex. On the other hand, larger antigens require to be transported by unknown cellular mechanisms. Antigens that are complexed with IgM, IgG, and complement might bind with Fc receptors or complement receptors on the surface of specialized macrophages, follicular DCs or B cells. Antigens presented on these surfaces can activate B cells through BCR cross-linking, expression of other interacting surface molecules, and cytokine secretion (Bonilla & Oettgen, 2010).

Two types of signals are required to activate B cells. Naïve B cells with IgM and IgD BCR isotype bind the antigen, but this is insufficient to initiate signal transduction. BCR cross-linking provides signal 1, which activates intracellular signaling pathways that enable the B cell to interact with T cell and thereby receive signal 2 (Rosenspire & Stemmer, 2010). B cells can function as APCs by expressing peptides via MHC II on their surface (see Figure 3.23). These peptides are produced from the degradation of antigens that were bound by BCRs and then internalized. These are then loaded into MHC II forming the peptide-MHC II complexes. When the B cell comes in contact with a Th cell with a T cell receptor (TCR) specific for such a peptide and with a CD4 molecule that recognizes the MHC II, the Th cell is activated to produce various cytokines. Then, the Th cell expresses CD40L that interacts with CD40 of the B cell. This interaction provides signal 2 to activate the B cell to express receptors that bind the various cytokines and cell surface molecules released by the Th cells. Consequently, the B cells are stimulated to multiplying the T cell area of the lymphoid tissue (Bonilla & Oettgen, 2010).

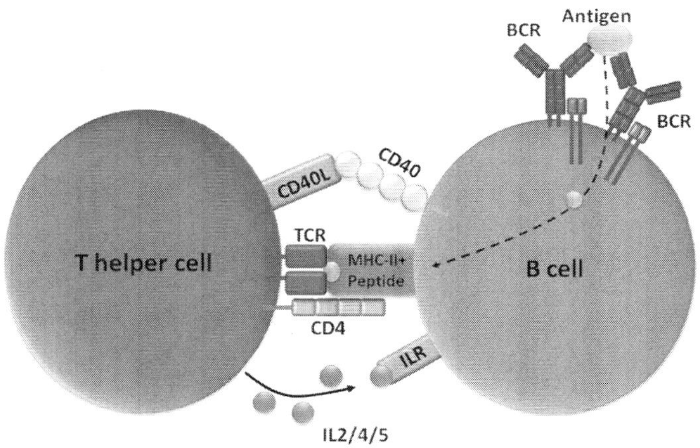

Figure 3.23. The T cell-dependent activation of B cells is initiated by a first signal generated by the cross-linking of B cell receptors (BCRs) upon antigen binding. The antigen is then ingested, processed, and presented by the B cell as peptide-major histocompatibility class II (MHC II) complexes, which are recognized by the T cell receptor (TCR)and CD4 of the helper T (Th) cell. The subsequent interaction of the B cell CD40 and the CD40L of the Th cell provides the second signal. Thereafter, the Th cell produces cytokines (e.g., IL2, IL4, IL5) to activate the B cells to proliferate. (Source: Altaileopard, Public Domain)

The activated B cells may relocate to the medullary cords and become short-lived plasma cells that secrete low-affinity immunoglobulins without somatic mutation to provide a rapid immune response and then undergo apoptosis. Alternatively, the activated B cells could enter, along with the attached Th cells, the marginal zone or follicles to establish a germinal center. This happens at least one week after contact with the antigen. In the spleen, the migration of B cells is regulated by chemokines like CCL21, CCL19, and CXCL13 that are produced by follicular DCs. In the germinal center, Tfh cells activate the B cells to proliferate and create the first part of the germinal center within the follicle. Somatic hypermutation also occurs in the loci of the variable regions of the immunoglobulin heavy and light chains. As a result, the B cells produce progeny cells with distinct receptors. Some of these receptors do not recognize the presented antigen while others manifest increased affinity. The B cells with increased affinity recognize and remain bound to antigen presenting DCs. At the same time, the B cells also undergo CSR, in which they switch from producing IgM and IgD to IgG, IgA or IgE. Somatic hypermutation and CSR result in affinity maturation that results in the production of B cells with a specific BCR isotype and with a higher affinity for the antigen. These B cells are then released from the secondary lymphoid organ into peripheral circulation (Bonilla & Oettgen, 2010; Cano & Lopera, 2013).

Some of these become long-lived plasma cells and go to the bone marrow while others remain in the lymphoid organ (see Figure 3.24). Each plasma cell can secrete several thousand molecules of immunoglobulin every minute, most of which are released into the bloodstream, and continue to do so for several days. The rate of secretion gradually decreases as the infection is cleared from the body. Nevertheless, some immunoglobulins persist for several months afterward. Tfh cells may also induce the high-affinity B cells to differentiate into memory cells. These cells multiply extensively but do not produce immunoglobulins. They remain in the tissues and in circulation for many months or even years. Upon activation by an antigen, these cells divide to form both plasma cells that secrete their specific immunoglobulin and another group of memory cells (Bonilla & Oettgen, 2010; Cano & Lopera, 2013; Britannica, 2021).

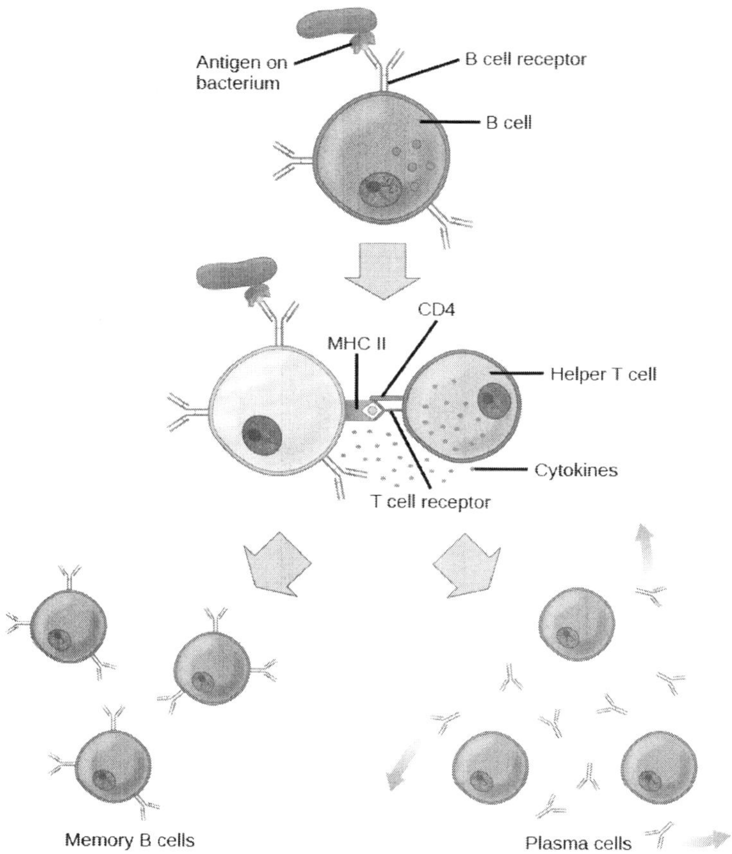

Figure 3.24. The T-dependent activation of B cells leads to the multiplication of B cells and their differentiation to plasma cells that secrete immunoglobulins or to memory cells. (Source: Charles Molnar & Jane Gair, Creative Commons License)

3.2.4 Immune Mechanisms by Immunoglobulins

Immunoglobulins may attach to pathogens or their toxins to render them harmless, a process called neutralization (see Figure 3.25). For instance, IgGs bind to diphtheria and tetanus toxins rendering them inactive. Immunoglobulins can also attach to receptors on the surface of pathogenic microorganisms preventing them from attaching themselves to the host cells. This makes the microorganisms unable to enter and infect the cells. Another way that bacteria are prevented from binding the host cells is through a process known as secretion blockade (see Figure 3.25). This occurs when the immunoglobulin binds to a secretion system and physically prevents protein secretion. In particular, the immunoglobulins bind to the translocon or needle tip proteins of the type III secretion system (T3SS). This creates a physical barrier that prevents the needle tip from correctly attaching to the translocon or the translocon from integrating into the host cell membranes (Hotinger & May,2020). Immunoglobulins may also bind the flagella of certain bacteria or protozoans rendering them immobile. Nevertheless, these protective mechanisms provided by immunoglobulins do not eliminate pathogens (Britannica, 2021).

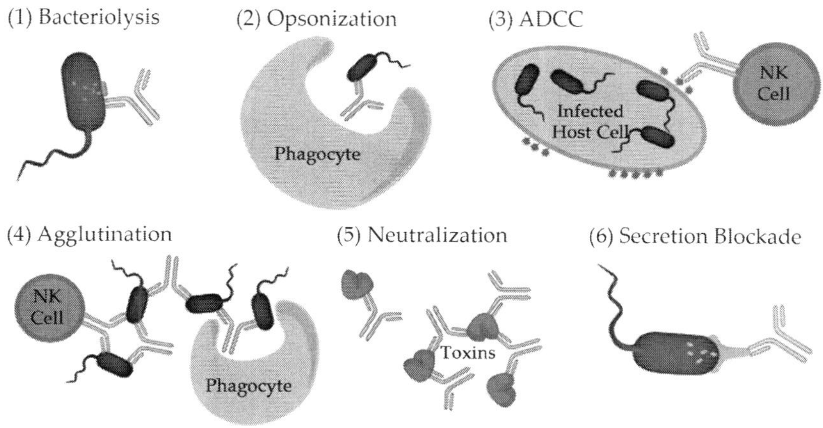

Figure 3.25. There are various mechanisms initiated by immunoglobulins to protect the host from pathogens. They can initiate the activation of the complement system resulting in (1) bacteriolysis or bursting of the bacterium or (2) opsonization that triggers phagocytes to ingest the pathogen. They also mediate the destruction of pathogenic cells by natural killer (NK) cells through a process called (3) antibody-dependent cell-mediated cytotoxicity (ADCC). (4) Agglutination or the aggregation of antigen-antibody complexes is essential in initiating opsonization and ADCC. Other mechanisms include the (5) neutralization of the pathogen or its toxin by the binding of the immunoglobulins and (6) secretion blockade that prevents type III secretion system (T3SS) proteins from being secreted. (Source: Julia A. Hotinger & Aaron E. May, Creative Commons License)

The destruction of pathogens is carried out via phagocytosis (see section 2.4.3), which is mediated by the complement system. Complement proteins do not interact with antigen binding sites. Nevertheless, immunoglobulins play a key role in mediating the activation of the complement system, specifically the classical complement pathway that is induced by antigen-

antigen complexes. The classical pathway is activated most effectively by IgM or IgG bound to antigens. In this pathway, C1 interacts with the Fc of the bound immunoglobulin. Once bound, C1 is cleaved and activated. It then triggers the sequential activation of complement proteins until C3 is activated. C3 is the most abundant and most biologically important component protein of the complement system. Upon activation, it is split into C3a and C3b, the large active fragment. C3b mediates the conversion of more C3 protein to C3b and initiates the alternative complement pathway. Moreover, it is involved in the lysis of the target cell by activating downstream reactions that result in the membrane attack complex. This is a ring-like structure made up of complement proteins C5 to C9. It inserts itself into the membrane of the pathogen to create holes and cause the cell contents to leak out, thereby killing bacteria (i.e., bacteriolysis) or viruses that lack protective coats (see Figure 3.25). C3b molecules are critical in opsonization, where several of them are deposited on the surface of the pathogen in order to tag them as targets of phagocytic cells. The receptors on the phagocytes recognize and bind the C3b proteins initiating phagocytosis (Britannica, 2021).

The small protein fragments released during the cleavage and activation of complement components can promote an inflammatory response by stimulating mast cells and basophils to release histamine and by attracting granulocytes and monocytes (Britannica, 2021).

Some cells with antigen-antibody complexes fail to activate the complement system. It maybe because the immunoglobulins are far apart or belong to an isotype that does not readily activate the complement system, such as IgA, IgD, and IgE. Other cells have tough outer membranes or membranes that are quickly repaired so that these are impermeable to activated complement molecules. Very large cells may also not be easily phagocytosed. In these cases, immunoglobulins stimulate killer cells to directly attack the cells that escape the defense mechanism provided by the complement system. This process is also called antibody-dependent cell-mediated cytotoxicity (ADCC) wherein cytotoxic T cells, NK cells, eosinophils or macrophages have receptors that bind to the Fc portion of IgGs that are bound to large pathogens or the cell infected with viruses or parasites (see Figure 3.25). Then these killer cells insert perforin protein into the target cell and secrete granzymes causing the target to swell and burst (Britannica, 2021).

ADCC is essential in providing protection from multicellular parasites such as hookworms and flatworms since they are too large to be ingested by phagocytes. Instead, they release so-called excretory/secretory (ES) antigens, like proteases, glycolytic enzymes, protease enzyme inhibitors, lectins, and antigens homologous to allergens, which are recognized and bound by DCs. The DCs then present the ES antigens to T cells inducing Th2 development, which is vital for parasite expulsion. Basophils and eosinophils can also sense ES antigens as well as process and present them to T cells. These cells also secrete IL4 which is essential in Th2 cell differentiation. It was also demonstrated that mast cells are involved in the development of Th2 cell-mediated responses. Th2 cells will then release certain cytokines like IL5, IL4, IL9, and IL13 during parasitic infection. In addition, immunoglobulins such as IgE, IgG, or IgG/IgA bind to the parasite and in turn, the Fc regions of these immunoglobulins are bound to the FcεRIon mast cells, basophils, macrophages, and eosinophils (Motran *et al.*, 2018) (see Figure 3.26). This interaction as well as the Th2 cell-mediated production of cytokines trigger the release of pharmacologically active granules from these cells resulting in a sudden increase in the permeability of local blood vessels, the adhesion and activation of platelets, the contraction of smooth muscles in the respiratory and digestive tracts, and the enhanced secretion of mucous by goblet cells which contribute to dislodge the parasites (Britannica, 2021). In addition, eosinophils release cytotoxic proteins, such as MBP1, MBP2, eosinophil peroxidase, ECP, and EDN, that digest the parasite's protective skin. Macrophages have been observed *in vitro* to be involved in the elimination of newly encysted juvenile worms of trematodes in a process mediated by nitric oxide. Nevertheless, neutrophils are predominant in the early response versus parasites. They are shown to be as effective as eosinophils in performing a cytotoxic attack against newborn larva of parasites via the involvement of H_2O_2 formation (Motran *et al.*, 2018).

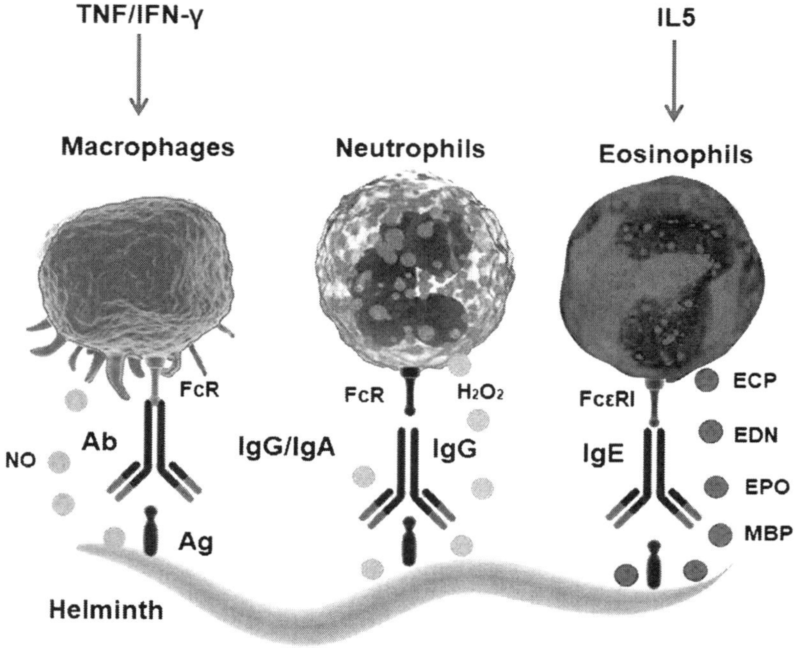

Figure 3.26. Antibody-dependent cell-mediated cytotoxicity (ADCC) is one mechanism by which helminth parasites are destroyed. Immunoglobulins interact with the parasite and their Fc is bound by Fc receptors found on macrophages, neutrophils, and eosinophils. This triggers the degranulation of these cells and the subsequent release of substances that lyse the parasite, such as nitric oxide (NO), hydrogen peroxide (H_2O_2), and toxic proteins like major basic protein (MBP), eosinophil peroxidase (EPO), eosinophil cationic protein (ECP), and eosinophil-derived neurotoxin (EDN). (Source: Claudia Cristina Motran *et al*., Creative Commons License)

Opsonization and ADCC are aided by agglutination (see Figure 3.25). This occurs when immunoglobulins bind multiple antigens forming large clumps that become easier to be targeted and ingested by phagocytes. A typical monomeric immunoglobulin has two antigen-binding (Fab) sites, but these may increase when multiple immunoglobulins or polymeric ones are involved. The formed aggregates may also activate NK cells and initiate ADCC as well as prevent the clumped bacteria from multiplying (Hotinger & May, 2020). Agglutination is also vital in a process called immune exclusion, wherein aggregated antigen-antibody complexes prevent pathogens from being adsorbed to mucosal surfaces so that they are swept out of the body (Britannica, 2021; de Sousa-Pereira & Woof, 2019).

3.2.5 Transfer of Maternal Immunoglobulins to Offspring

An offspring inside the uterus develops its own innate immune system but has no opportunity to develop protective immunoglobulins unless it is infected while in the uterus, which is a very rare occurrence. In addition, the placenta prevents maternal lymphocytes from crossing into the uterus to prevent the mother's immune system from recognizing the fetal tissues as foreign antigens and rejecting them (Britannica, 2021). Thus, newborns are not yet equipped with a fully developed immune system after birth.

Still, a newborn has a measure of adaptive immunity when it is born and exposed to antigens in the external environment because maternal immunoglobulins, which are mostly of the IgG isotype, may be transferred across the placenta to the fetal circulation (see Figure 3.27). This starts at the early stages of pregnancy, although at low efficacy, then reaches its peak during the second trimester. It is mediated by the neonatal Fc receptor (FcRn), which transports the maternal immunoglobulin in a pH-dependent manner. IgG in the maternal blood is taken in by the placental syncytiotrophoblasts via endocytosis at their apical side. FcRn is expressed in these cells and binds the Fc fragment of maternal IgGs in the acidic environment (i.e., pH=6) of the endosomes. When the endosome fuses with the plasma membrane at the basolateral side, the IgG-FcRn complex is exposed to extracellular fluid with neutral pH (i.e., pH=7-7.4). The change in pH decreases the affinity of the receptor to the immunoglobulin so that the IgG is released and then diffuses into the fetal bloodstream (Britannica, 2021; Schroeder & Cavacini, 2010; Albrecht & Arck, 2020). Not all IgG subclasses are transferred with equal efficiency. FcRn mainly transports IgG1 and then transports IgG4, IgG3, and IgG2 with decreasing efficacy, respectively. Therefore, newborns might lack specific immunity toward the target pathogens of IgG2, such as *Haemophilus influenzae* type b or *Neisseria meningitidis* (Albrecht & Arck, 2020).

Passively transferred maternal IgGs have temporary effects. Their level in the blood becomes diluted as the child grows and they also gradually decline due to normal metabolic breakdown. Since acquired immunity develops gradually and thus may not immediately replenish the depleting maternal immunoglobulins, the child is more susceptible to infection during the early stages of growth after birth than immediately after birth (Britannica, 2021; Niewiesk, 2014).

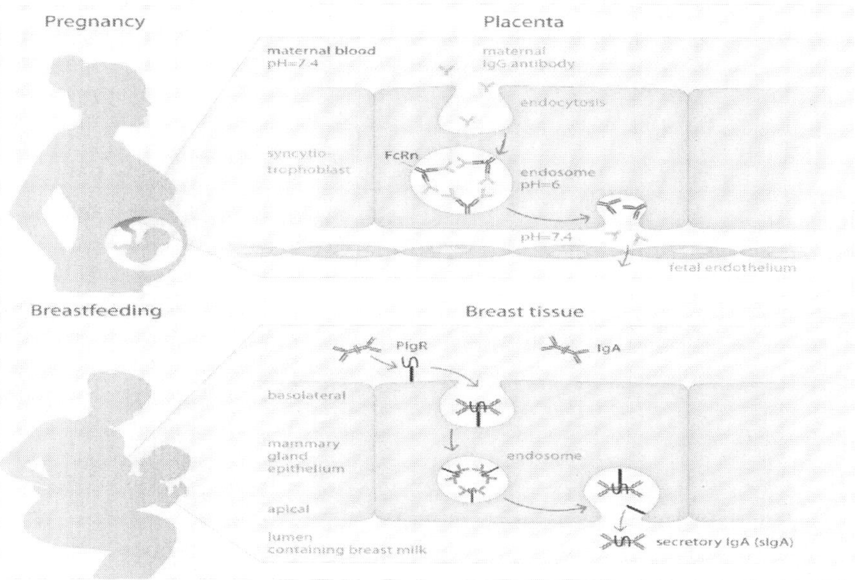

Figure 3.27. Maternal IgG is taken up into the syncytiotrophoblast cell of the placenta via endocytosis. The neonatal Fc receptor (FcRn) at the inner membrane of the acidic endosome binds two IgG molecules. Upon fusion of the endosome with the basolateral membrane, FcRn releases the IgGs due to the increased pH of the extracellular fluid at the fetal side of the placenta. The FcRn can then be recycled to perform another transport cycle. (Source: Marie Albrecht & Petra Clara Arck, Creative Commons License)

On the other hand, breastmilk is rich in sIgA with the colostrum observed to have the highest concentration within 24 hours after birth (Britannica, 2021; Niewiesk, 2014). Breastmilk may also contain some maternal immune cells like IgG-producing memory B cells and CD4T cells. Unlike maternal IgGs, maternal IgA is continuously replenished through the breastmilk taken in by the infant. The dimeric IgA molecules are produced by plasma cells in the mammary gland and in other tissues associated with mucosal surfaces. IgA molecules from the connective tissue are transported via transcytosis across the epithelial cells of the mammary acini to the breastmilk (see Figure 3.28). This involves the pIgR, expressed on the basolateral surface of the mammary cell, which binds the joining chain of the dimeric IgA. The resulting pIgR-IgA complex is internalized via endocytosis. The endosome then fuses with the apical membrane and the pIgR releases the sIgA into the breastmilk with the SC of the pIgR still bound to the immunoglobulin. When the infant takes in breastmilk, sIgA is not absorbed by the intestine, but it coats the mucosal surface of the gastrointestinal tract to protect it from pathogens (Albrecht & Arck, 2020).

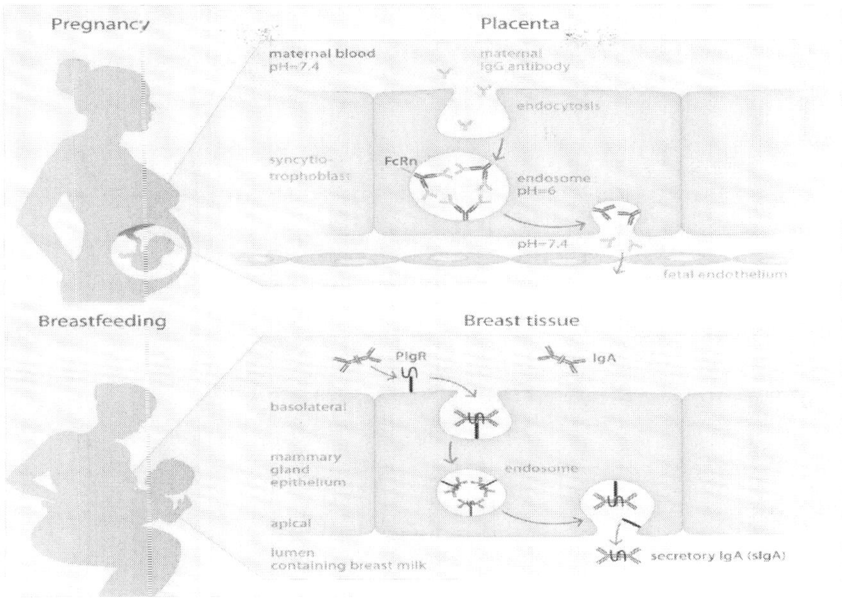

Figure 3.28. The polymeric Ig receptor (pIgR) binds the joining chain of the dimeric IgA molecule at the basolateral side of the mammary epithelial cell. The pIgR-IgA complex is then taken in by endocytosis. When the endosome fuses with the apical membrane of the cell, the pIgR releases the secretory form of the IgA (sIgA) into the lumen that contains the breastmilk. (Source: Marie Albrecht & Petra Clara Arck, Creative Commons License)

The specificity of the immunoglobulins received by the newborn depends on the particular pathogen against whom the mother has immunity. Therefore, a prospective mother may be asked to receive immunizations against certain diseases, such as tetanus or influenza (Britannica, 2021). One study demonstrated the transfer of influenza-specific IgA to the infant from the mother who was vaccinated against influenza during her pregnancy. Also, mothers who had respiratory tract infections during pregnancy had higher levels of IgA1 in their colostrum. This subclass is known to be mainly present in the respiratory tract. Meanwhile, pregnant women with gastrointestinal infections had a higher proportion of IgA2, mainly found in the intestines, in the colostrum (Albrecht & Arck, 2020).

3.2.6. Hypersensitivity Reactions

Some immunoglobulins like IgE, IgM, and IgG are involved in hypersensitivity reactions, which are undesirable or inappropriate reactions produced by the immune system against an antigen or allergen. There are four types of hypersensitivity reactions with the first three considered to be immediate hypersensitivity reactions (IHR) because they occur within 24 hours (Warrington *et al.*, 2011; Vaillant *et al.*, 2021).

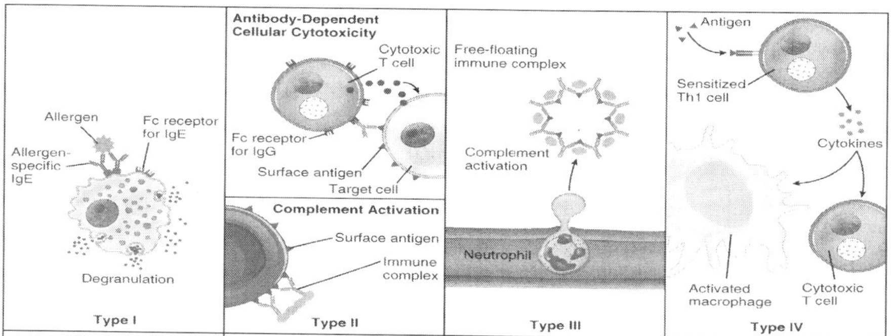

IgE-Mediated Hypersensitivity	IgG-Mediated Cytotoxic Hypersensitivity	Immune Complex-Mediated Hypersensitivity	Cell-Mediated Hypersensitivity
IgE is bound to mast cells via its Fc portion. When an allergen binds to these antibodies, crosslinking of IgE induces degranulation.	Cells are destroyed by bound antibody, either by activation of complement or by a cytotoxic T cell with an Fc receptor for the antibody (ADCC)	Antigen–antibody complexes are deposited in tissues, causing activation of complement, which attracts neutrophils to the site	Th1 cells secrete cytokines, which activate macrophages and cytotoxic T cells and can cause macrophage accumulation at the site
Causes localized and systemic anaphylaxis, seasonal allergies including hay fever, food allergies such as those to shellfish and peanuts, hives, and eczema	Red blood cells destroyed by complement and antibody during a transfusion of mismatched blood type or during erythroblastosis fetalis	Most common forms of immune complex disease are seen in glomerulonephritis, rheumatoid arthritis, and systemic lupus erythematosus	Most common forms are contact dermatitis, tuberculin reaction, autoimmune diseases such as diabetes mellitus type I, multiple sclerosis, and rheumatoid arthritis

Figure 3.29. Among the four types of hypersensitivity reactions, the first three involve immunoglobulins while type IV is T cell-mediated. Type I is mediated by IgE that leads to degranulation of mast cells and basophils so that they release molecules that stimulate the allergic response. Type II involves IgG and ,in some cases, IgMthat bind to self-antigens activating antibody-dependent cellular cytotoxicity or the complement system. Type III is characterized by the formation of immune complexes consisting of antigens bound by immunoglobulins that eventually activates the complement system and neutrophil influx. Type IV response is initiated by a sensitized helper T cell that produces cytokines resulting to activation of macrophages or cytotoxic T cells (Source: J. Gordon Betts *et al.*, Creative Commons License)

The most common type of this is type I hypersensitivity or anaphylactic response, which is an allergic reaction triggered by re-exposure to an allergen like cat, dog or horse epithelium, pollen, house dust mites, and molds as well as food allergens (see Figure 3.29). Such allergens are proteins with a molecular weight ranging from 10-40 kDa. They stimulate the production and the secretion of IgE upon first exposure to them. IgEs bind to Fc receptors on mast cells and basophils causing the cross-linking of IgE and the cells to be sensitized. When these immunoglobulins are exposed to the same allergen, the sensitized cells are degranulated, and they secrete mediators, such as histamine, leukotriene, and prostaglandin that cause vasodilation, inflammation, and smooth-muscle contraction of the surrounding tissue (Warrington *et al.*, 2011; Vaillant *et al.*, 2021).

Type II hypersensitivity occurs when IgG and IgM bind to the person's own cell surface receptors and extracellular matrix proteins to form complexes that activate the complement system or ADCC (see Figure 3.29). This type of reaction occurs rarely and takes 2 to 4 hours to develop. Once the complement system is activated, opsonization, cell agglutination, cell lysis and death ensue. The immunoglobulin bound to the target cell may also interact with a cytotoxic T cell via its Fc fragment to activate ADCC. Some examples of these reactions are hemolytic transfusion reactions, erythroblastosis faetalis, Goodpasture's syndrome, and autoimmune anemias (Warrington *et al.*, 2011).

When IgG and IgM bind to soluble proteins to form antigen-antibody aggregates called immune complexes, this results in type III hypersensitivity reactions (see Figure 3.29). These immune complexes are deposited in tissues so that they accumulate over time, i.e., hours to weeks. This leads to complement activation, inflammation, neutrophil influx, and mast cell degranulation resulting in tissue damage. Examples of this type of reaction include systemic lupus erythematosus (SLE), serum sickness, and reactive arthritis (Warrington *et al.*, 2011; Vaillant *et al.*, 2021).

Type IV hypersensitivity reactions are not dependent on immunoglobulins, but rather cell-mediated (see Figure 3.29). They are the second most common type of hypersensitivity reaction and usually take 2 or more days to develop. These are caused by the overstimulation of Th cells resulting to the release of cytokines that activate monocytes or macrophages and cytotoxic T cells. This leads to inflammation, cell death, and tissue damage (Warrington *et al.*, 2011).

Table 3.3 presents the different types of hypersensitivity reactions.

Table 3.3. Types of Hypersensitivity Reactions

Type	Alternate name	Mediators	Examples
I	Allergy or Anaphylactic response	IgE	Atopy Asthma Anaphylactic shock Allergic conjunctivitis Allergic rhinitis Angioedema Drug allergy Food allergy
II	Cytotoxic, antibody-dependent	IgG, IgM	Erythroblastosis faetalis Hemolytic transfusion reactions Goodpasture's syndrome Autoimmune anemias Myasthenia gravis Thrombocytopenias
III	Immune complex disease	IgG, IgM, complement proteins, aggregation of antigens	Systemic lupus erythematosus Serum sickness Reactive arthritis Arthrus reaction
IV	Delayed-type hypersensitivity, cell-mediated, antibody-independent	T cells, monocytes, macrophages	Contact dermatitis Tuberculosis Chronic transplant rejection

Source: Warrington *et al.* (2011). *Allergy, Asthma & Clinical Immunology* 7: S1

3.3 CELL-MEDIATED IMMUNE RESPONSE

Cell-mediated immunity is attributed to T cells and does not involve immunoglobulins. It involves the activation of macrophages and NK cells to destroy intracellular pathogens, the production of antigen-specific CD8 T cells, and the production of cytokines that facilitate the function of both innate and adaptive immune cells. This type of immune response is the principal mechanism for eliminating virus-infected cells. It is also vital in providing protection from fungi, protozoa, intracellular bacteria, and cancers as well as in facilitating rejection of transplanted organs (Bhagavan &Ha, 2011).

3.3.1 The T Cell Receptors

T cell receptors (TCRs) recognize antigen-MHC complexes (see Figure 3.30a). These are heterodimers made up of two polypeptide chains and embedded on the T cell membrane. A T cell usually has about 20, 000 TCRs on its surface. There are two types of TCR. The most common one is called alpha-beta (α-β) since it is made up of one alpha chain and a beta chain. It is expressed in the majority of T cells. A less common type is the gamma-delta (γ-δ) receptor composed of the gamma and delta chains, and it occurs in peripheral blood T cells (Britannica, 2021; Rosati *et al.*, 2017).

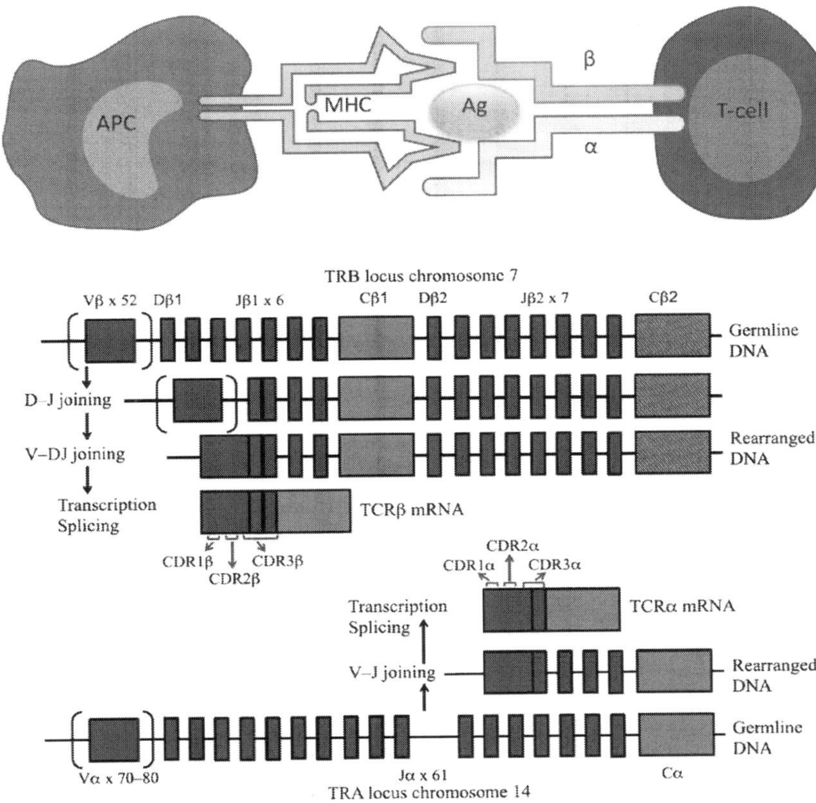

Figure 3.30: (a) A T cell receptor (TCR) interacts with an antigen bound by a major histocompatibility complex (MHC) molecule on the surface of an antigen presenting cell (APC). (b) The TCR α chain (TRA) locus on chromosome 14 has an array of several variable (V) and joining (J) gene segments as well as a constant (C) gene segment. The TCR β chain (TRB) locus on chromosome 7 has the same kinds of gene segments with an additional diversity (D) segments.

During T cell development, the V(D)J segments in the germline DNA are rearranged to form the final rearranged DNA sequence. Following transcription, the sequence between the recombined V(D)J regions and the C region is spliced out. The complementarity-determining region (CDR) 1 and CDR2 are encoded within the germline V gene segment, while the CDR3 is coded for by the V(D)J junction. (Source: Elisa Rosati *et al.*, Creative Commons License)

Immunoglobulins and TCR genes have a similar organization with the TCR loci containing arrays of V, D, and J gene segments as well as C gene segments. All TCR loci have V and J segments, but only the β and δ loci have D segments. The α and δ loci are located on chromosome 14, while the β and γ loci are on chromosome 7. The V(D)J recombinase catalyzes the splicing and joining together of a V, a D (for β and δ chains), and a J segment to form the V region locus (see Figure 3.30b). Then the recombination of the V region with the C gene segment produces the TCR chain transcript. The process of recombination is similar to what happens during the generation of immunoglobulins that is discussed in section 3.2.1.2.The V(D)J recombinase cleaves the germline DNA at the RSS which is at the border of the V, D, and J segments. This produces DSBs and then NHEJ repair of the DNA breaks results in the joining of the V, D, and J segments. In the β chain locus, the first round of somatic recombination joins the D and J segments together. A second round of excision and ligation brings together any one of the 70 V gene segments with the DJ segment. The large number of possible combinations of V, D, and J segments generate TCRs with different binding specificities. Junctional diversity also increases further the variability of TCRs (Bonilla & Oettgen, 2010). The repertoire of the TCR β chain alone is estimated to range from 1 x 10^6 to 3 X 10^6 (Boegel *et al.*, 2019). On the other hand, the diversity of TCR dimers is approximately 10^{13}-10^{15} (Rosati *et al.*, 2017).

If the somatic recombination and subsequent diversification do not produce stop codons, the resulting transcript codes for a full-length TCR protein. Such viable rearrangement should take place sequentially in order for a T cell to express two TCR chains, i.e., αβ or γδ. The structures of the resulting proteins are also similar to immunoglobulins. Each TCR chain contains two folded domains, a constant and variable one (see Figure 3.31). Just like in an immunoglobulin, the variable domains of the chain form an antigen-binding site. However, the TCR has a single antigen-binding site instead of two (Britannica, 2001; Bonilla & Oettgen, 2010). A TCR chain contains three CDRs (i.e., CDR1, CDR2, and CDR3). CDR1 and CDR2 are required for the interaction of the TCR with the MHC complex. These are encoded by V genes (see Figure 3.30b). CDR3 is the region that comes in direct contact with the peptide antigen and thus, it determines the antigen binding specificity of the TCR. It undergoes a significant change in conformation assuming a diagonal position that enables it to bind the peptide-MHC complex. It is encoded by the junctional region between the V and J or D and J gene segments making it highly variable (Rosati *et al.*, 2017; Tai *et al.*, 2018). CDRs can be re-edited via TCR revision, which changes the antigen specificity of the TCR. This is initiated by reactivation of the recombinase complex that triggers DNA recombination. This forms TCRs that can tolerate self-antigens but are still capable of providing protection against pathogens (Bio-Rad, 2021).

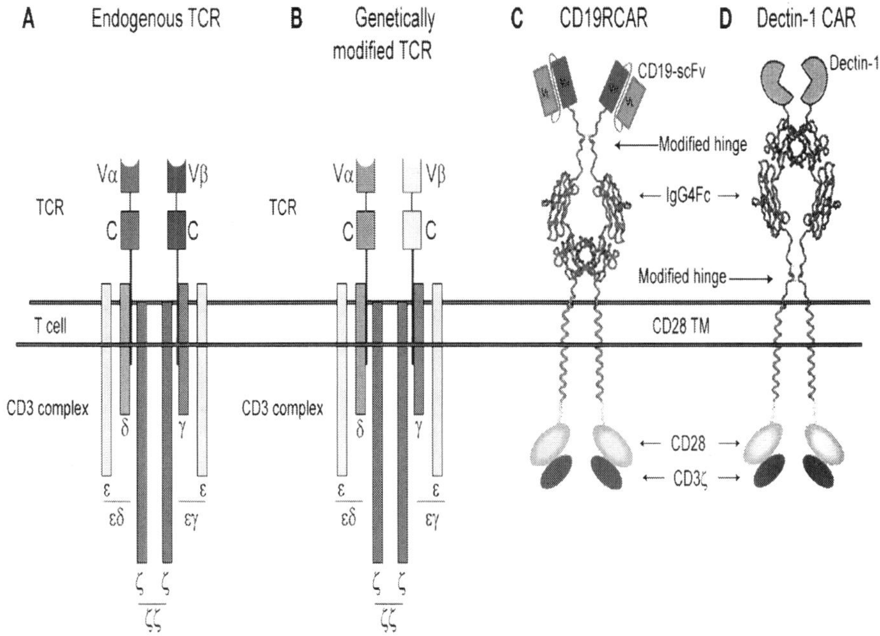

Figure 3.31. The T cell receptor (TCR) is a heterodimer (αβ) located on the T cell membrane with a variable (V) and a constant (C) domain in each chain. It is associated with a complex of CD3 proteins forming three dimers, namely epsilon-delta (εδ), epsilon-gamma (εγ), and zeta-zeta (ζζ). (Source: Pappanaicken R. Kumaresan *et al.*, Creative Commons License)

The TCR chains are assembled at the cell surface and form a complex withCD3 proteins, which are vital for activating T cells (see Figure 3.31). CD3 proteins are signal transducers that mediate the conversion of extracellular binding of ligand and receptor into intracellular signals. These signals are then transmitted to the nucleus via the cytoplasmic ITAMs of CD3 (Britannica, 2001; Bonilla & Oettgen, 2010). There are six CD3 chains per TCR and these form three dimers, i.e., εδ, εγ, and zeta-zeta(ζζ). Thus, the TCR complex is made up of spatially separated ligand-binding and signal transduction subunits. These subunits assemble together via the interaction of the extracellular immunoglobulin domains of the CD3 subunits and the TCR αβ chains and the presence of nine conserved amino acid residues in the transmembrane regions (Mingueneau, 2016).

Unlike immunoglobulins, TCRs cannot bind to free-floating antigens. They bind fragments of foreign proteins that are ingested, partially digested, and displayed on cell surfaces by the MHC molecules. Therefore, intracellular pathogens that are out of reach of immunoglobulins are still susceptible to being destroyed by T cells (Britannica, 2001). TCR αβ typically recognizes peptide-MHC complexes. It may also bind non-peptidic antigens that are presented by the unconventional MHC Ib molecules. TCR αβ expressed by NKT cells interacts with lipid antigens that are bound to the MHC I-related CD1 surface receptor. CD1 has 30% homology with MHC I and there are five types (i.e., CD1a, CD1b, CD1c, CD1d, and CD1e). Meanwhile, some γδ T cells recognize peptides, but most bind atypical antigens that may or may not be associated with an antigen-presenting molecule (Schroeder *et al.*, 2019; Tai *et al.*, 2018).

TCRs also manifest cross-reactivity, which is attributed to the flexibility in their binding to peptide-MHC complexes. As a result, fewer T cells are needed to detect an antigen. An antigen may be recognized by various TCRs in several T cells, which makes it difficult for the pathogen to evade being detected by the immune system (Bio-Rad, 2021).

3.3.1.1. T cell differentiation

The expression of TCR chains on the cell surface signals the development of precursor T cells into double-positive T cells, which are found in the cortex of the thymus. These are called such because they express both CD4 and CD8. These undergo further differentiation to form single-positive T cells, which are located in the thymic medulla. The TCR of the double-positive T cells binds with low avidity to MHC complexed with self-peptides on the thymus epithelium. Those that fail to interact with the self-peptide-MHC complexes are eliminated. This process of favoring the development of T cells with functional TCRs is called **positive selection**. On the other hand, the double-positive T cells with TCR that binds with very high avidity to the self-peptide-MHC complex are said to be **negatively selected** through a process facilitated by an autoimmune regulator. This ensures that only those with TCRs of intermediate affinity remain and autoreactive T cells are eliminated (Bonilla & Oettgen, 2010).

Double-positive T cells that survive the positive or negative selection mature into CD4 single-positive T cells when they interact with MHC II molecules on the thymus epithelium. Those that interact with MHC I molecules become CD8 T cells. These two types of single-positive T cells then leave the thymic medulla and enter the bloodstream as fully differentiated but antigen naïve T cells (Bonilla & Oettgen, 2010).

Along with the classical CD4 and CD8 T cells, some nonconventional types of T cells are also produced. These include FOXP3 CD4 CD25 natural Tregulatory (nTreg) cells, the CD1d-reactive natural killer T (NKT) cells, the MHC1b CD8 T cells, and the major histocompatibility molecule-related 1 (MR1)- restricted mucosa-associated invariant T cells (Luckheeram *et al.*, 2012).

The naïve T cells, particularly CD4 T cells, enter the lymph nodes via specialized blood vessels called high endothelial venules (HEVs). The L-selectin on the T cells binds to vascular addressin on the HEVs to enable the specific homing of T cells into the lymphoid tissue in response to chemokines. In addition, the HEVs and stromal cells in the lymphoid tissue produce CCL21 that binds to its CCR7 receptor on naïve T cells. This promotes the recruitment of T cells into the lymph node where they are activated (Camiolo *et al.*, 2020).

3.3.2 T Lymphocyte Activation

Two signals are necessary to activate T lymphocytes or T cells. The activation of T cells is initiated by the first signal, which is when the TCR and associated proteins recognize a peptide-MHC complex presented by an APC (see Figure 3.32). This leads to rapid aggregation of the TCR and its associated molecules at the physical interface between the T cell and APC to form an immunologic synapse, also called the supramolecular activation complex. At the T cell side, the synapse consists of a central cluster of CD3 molecules (i.e., γ, ε, and ζ) and TCR (α and β) that bind specifically to the peptide-MHC complex. The same side also contains the CD4 or CD8 molecule, whose extracellular domain binds the nonpolymorphic portions of MHC II or MHC I, respectively and thus stabilizes the T cell-APC interaction (see Figure 3.33). The synapse is further stabilized by integrins. The intracellular tails of CD4 and CD8 interact with Lck, the kinase that initiates TCR signaling. Thus, the formation of the synapse sets off and maintains a signaling cascade that induces NF-κB- and MAPK-mediated expression of genes as well as cytoskeleton reorganization that are critical for the function of activated T cells (Bonilla & Oettgen, 2010).

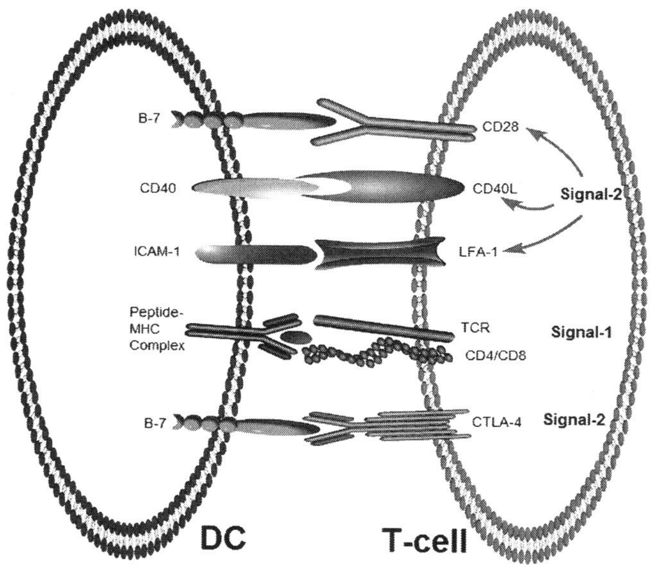

Figure 3.32. Two signals are required in order to fully activate T cells. Signal 1 is provided by the interaction of TCR and the peptide-major histocompatibility complex (MHC). The binding of CD4 or CD8 with the MHC stabilizes such interaction. Signal 2 is provided by the binding of co-stimulatory receptors (e.g., CD28, CD40L, LFA-1) with their corresponding ligands. An inhibitory signal may also be transmitted via the interaction of CTLA-4 with B7. (Source: Yu Tai *et al.*, Creative Commons License)

Apart from signaling via TCR, a second signal is essential to completely activate the T cells (see Figure 3.32). This involves other cell surface receptors on the T cell membrane, mainly CD28 but may also include CD2, CD5, CD30, LFA-1, inducible costimulatory (ICOS) molecule, and members of the TNF receptor family (e.g., CD27, OX-40, and 4-1BB) (see Figure 3.33). The binding of these co-stimulatory receptors to their ligands amplifies TCR signaling and provides complementary signals that are delivered to the T cell via the ITAM of the cytoplasmic domains of the receptors. The second signal sustains the activation cascade long enough to enhance survival and rapid proliferation, support cytokine production to promote effector differentiation, and generate memory cells (Bio-Rad, 2021; Chandler *et al.*, 2020). The two-signal mechanism ensures that T cells do not respond to self-antigens since those activated with just one signal would respond to any peptide that fit their TCRs without differentiating self- from non-self-antigens (Birdsall & Casadevall, 2020). If a T cell receives signal 1 but not signal 2, it may undergo apoptosis or it may be altered so that it can no longer be activated even if it receives both signals at a later time (Alberts *et al.*, 2002).

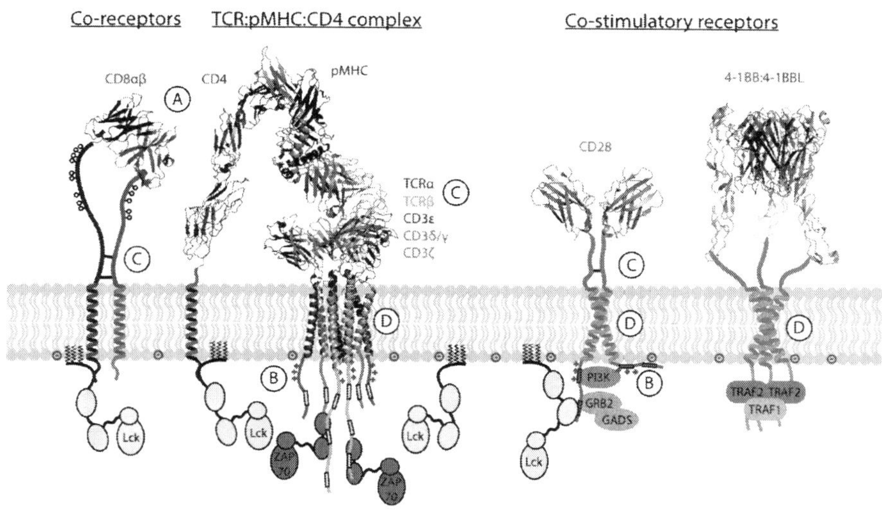

Figure 3.33. The CD4 or CD8 serve as co-receptors on the T cell membrane. They have extracellular domains that engage the major histocompatibility complex (MHC) to stabilize the peptide-MHC interaction. Their cytoplasmic tails associate with the kinase Lck. Also, there are co-stimulatory receptors on the T cell surface, such as CD28 and 4-1BB. (Nicholas J. Chandler *et al.*, Creative Commons License)

Among the APCs, DCs are considered to be the most significant since they have enhanced ability to activate naïve T cells and they express the ligands of co-stimulatory receptors on the T cells (Luchkeeram *et al.*, 2012). However, they, along with monocytes, engage with antigens on a random basis, On the other hand, B cells capture specific antigens via the IgMs on their membranes. Interaction with the antigen leads to cross-linking of the IgMs, and this activates the B cell to express more MHC II molecules. Therefore, the B cells interact with the T cells with the appropriate antigen specificity. B cells also produce costimulatory molecules to activate the CD28 co-receptor (Birdsall & Casadevall, 2020).

3.3.2.1. CD4 T cell activation

The differentiation of naïve CD4 T cells into helper T (Th) cells occurs in the peripheral lymphoid tissues where the TCRs of naïve T cells engage with APCs that display non-self peptide-MHCII complexes(see Figure 3.34). This interaction provides the first signal for T cell activation. The second signal is provided by the interaction of co-stimulatory molecules like B7 and CD40 or certain cytokines with their respective receptors on the CD4 T cell. B7 is found on the APC surface, and it has two homologous subsets, namely B7-1 (or CD80) and B7-2 (or CD86). It binds CD28 that is constitutively expressed on the CD4 T cell surface leading to increased production of IL2 and other cytokines, increased expression of cytokine receptors, improved cell survival, and enhanced T cell proliferation. Meanwhile, CD40 on B cells and other APCs interacts with CD40L on the CD4 T cell. This interaction upregulates the expression of B7 and other adhesion molecules as well as cytokine production by the presenting cell (Bhagavan & Ha, 2011; Britannica, 2021; Broaddus, 2022).

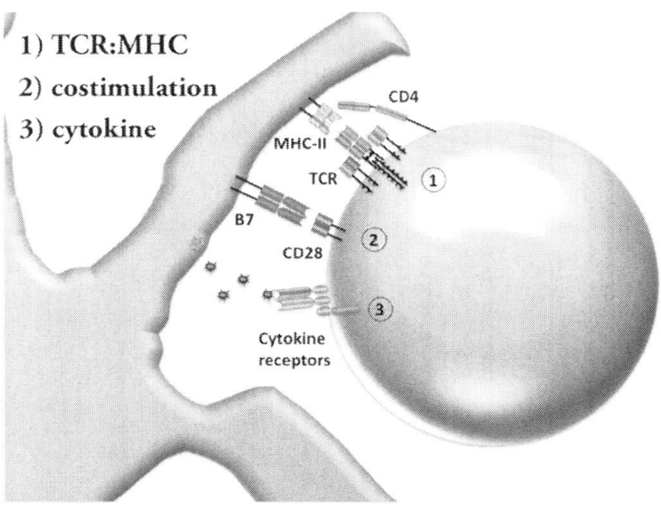

Figure 3.34. Naïve CD4 T cells are activated by two signals. The first one is produced when the T cell receptor (TCR) binds a peptide presented by the major histocompatibility complex II (MHC II) molecule on the surface of an

antigen presenting cell (APC). The CD4 co-receptor stabilizes the TCR-MHC II interaction. The second signal is provided by the interaction of co-stimulatory molecules on the APC like B7.1 and B7.2 with CD28 on the T cell. Finally, cytokines from macrophages and dendritic cells stimulate the CD4 T cell to differentiate into helper T (Th) cells. (Source: Hope O'Donnell &Stephen J. McSorley, Creative Commons License)

Accumulated pathogens inside the vesicles of macrophages and DCs as well as the cytokines IFNγ and IL12 stimulate the differentiation of naïve CD4 T cells into Th1 cells (see Figure 3.35). Th1 differentiation is mainly regulated by the T-box transcription factor (T-bet), which significantly enhances IFNγ production and is involved in suppressing Th2 and Th17 development (Luckheeram *et al.*, 2012).

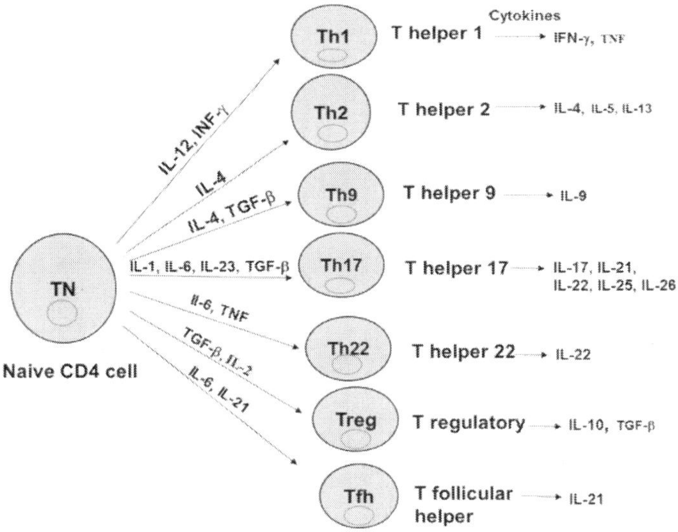

Figure 3.35. Naïve CD4 T cells differentiate into various types of helper T (Th of Tfh) cells or into regulatory T (Treg) cells through the action of a different combination of cytokines. In turn, each type of activated CD4 T cells produces a distinct set of cytokines. (Source: Vita Golubovskaya & Lijun Wu, Creative Commons License)

On the other hand, Th2 cell-mediated immune response is stimulated by extracellular pathogens and allergens as well as the cytokines IL2 and IL4 (see Figure 3.35). The major transcription factor involved in Th2 cell differentiation is the IL4-induced STAT6, which promotes the expression of GATA-binding protein 3 (GATA3). GATA-3 facilitates Th2 cytokine production, the selective proliferation of Th2 cells, and inhibition of Th1 cell differentiation. Other cytokines involved in Th2 cell differentiation include IL6 and IL21 (Luckheeram *et al.*, 2012).

TGF-β diverts the differentiation of Th2 cells into the development of Th9 cells (see Figure 3.35). Also, TGF-β along with IL4 directly induces Th9 cell differentiation with IRF4 also playing an important role. Meanwhile, Th17 cell differentiation is mainly regulated by retinoic acid receptor-related orphan receptor gamma-T (ROR γt) functioning synergistically with RORα. There are three stages during Th17 formation with TGF-β and IL6involved in the differentiation step. IL21 facilitates the self-amplification stage and IL23 is involved in the stabilization stage (see Figure 3.35). Self-amplification is vital in providing a robust immune response, whereas stabilization ensures expansion and maintenance of the Th17 pool (Luckheeram *et al.*, 2012). The development of Th22 is facilitated by TNF and IL6(Golubovskaya & Wu, 2016) (see Figure 3.35).

The development of Treg cells is mediated by TGF-β and IL2 (Golubovskaya & Wu, 2016) (see Figure 3.35). They may be natural (nTreg) or induced (iTreg). The former already constitutively expresses the forkhead transcription factor FOXP3and hasCD25 when it leaves the thymus, while the latter develops in the peripheral lymphoid tissues and expresses FOXP3and CD25 after exposure to antigens. TGF-β initiates iTreg cell lineage commitment by inducing the expression of FOXP3, which is the major transcription factor involved in iTreg cell differentiation. Tfh cells, on the other hand, are found in the spleen and lymph nodes, particularly at the germinal center. Their differentiation is mediated mainly by IL6 and IL21, which activate the transcription factor STAT3(Luckheeram *et al.*, 2012) (see Figure 3.35). They are sometimes classified as a type of memory cell due to their long life span, which is maintained by a persistent antigen (Jaigirdar & MacLeod, 2015).

3.3.2.2 CD8 T cell activation

In secondary lymphoid organs, the TCRs of naïve CD8 cytotoxic T cells, also known as the killer T cells or cytotoxic T lymphocytes (CTLs), bind the peptide-MHC I complex on APCs (see Figure 3.36). Then the binding of CD28 on the CD8 T cell to CD80 or CD86 on the APC provides costimulatory signals. These signals along with cytokines like IL12 andIFNIactivate the CD8 T cells to proliferate. Moreover, these signals activate several biochemical pathways that significantly alter the metabolism of the T cells such that they have increased uptake of glucose, amino acids, and iron, and they switch from performing oxidative phosphorylation to aerobic glycolysis, perhaps to provide the energy needed for producing nucleic acids, proteins, and lipids. All these increase the rate of CD8 T cell division by as much as 500, 000 times (Zhang & Bevan, 2011).

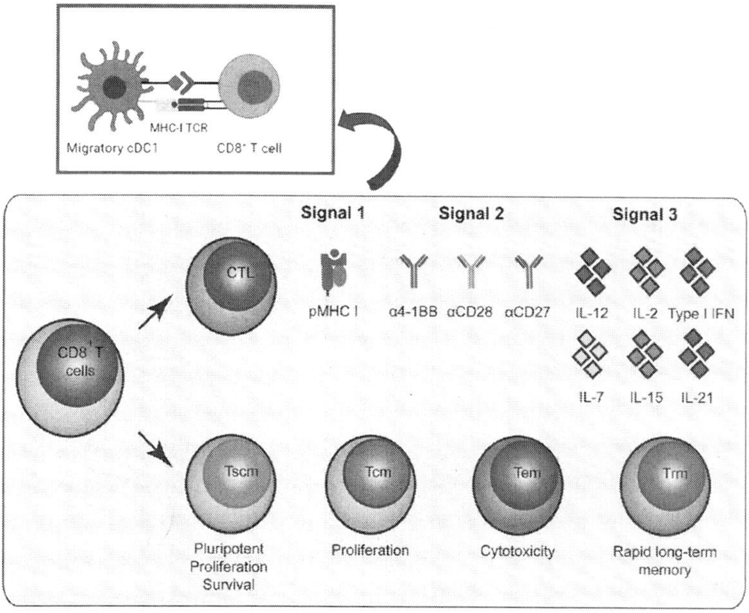

Figure 3.36. The first signal that activates CD8 T cells is provided by their recognition of peptide-major histocompatibility complex I (pMHC I) complex on an antigen presenting cell like a migratory conventional dendritic cell (see inset). A complementary signal is provided by co-stimulatory molecules like 4-1BB, CD28, and CD27. These two signals along with certain cytokines stimulate the CD8 T cells to proliferate and enhance their cytotoxic functions. After proliferation, many of the T cells undergo apoptosis leaving a few that differentiates into different types of memory cells. (Source: Marjorie Schluck *et al.*, Creative commons license; inset: Julie Busselaar *et al.*, Creative Commons License)

However, the resulting CD8 T cells have weak effector functions. Their effector function can be improved through the action of cytokines including IL2 and IFNγ that are produced by CD4 Th1 cells. These cytokines may directly enhance the function of the naïve CD8 cytotoxic T cell or may indirectly contribute to it by improving the ability of the interacting APC (Bhagavan & Ha, 2011).

CD8 T cells may also differentiate into Treg cells that differ in terms of their surface marker profile. There are those expressing CD25 and FOXP3 similar to CD4 nTreg cells. Others express CD122 and PD1 with the latter protein being responsible for the suppression of autoimmunity and for facilitating tolerance to antigens and thereby, inhibiting transplant rejection. CD8 Treg cells also secrete immunosuppressive cytokines (IL4, IL10) and TGFβ and regulate homeostasis by inducing apoptosis of activated T cells via Fas-FasL interaction (Ratajczak *et al*., 2018).

After proliferation or clonal expansion, the CD8 T cells undergo the **contraction phase**, in which a large number of them experience apoptosis. The remaining antigen-specific CD8 T cells have decreased effector function and become memory T cells, which become reactivated when exposed to the same antigen for the second time. This memory phase may last throughout a person's lifetime (see Figure 3.35) (Perdomo-Celis *et al*., 2019).

3.3.2.3 Inhibition of T cell activity

After the T cells have been activated, they express cytotoxic T lymphocyte antigen-4 (CTLA4) on their surface. This binds more strongly than CD28 to B7 on the APC. This interaction provides an inhibitory signal to suppress the T cell response. It appears to be essential in the development of anergy and the generation of peripheral tolerance. Programmed cell death 1 (PD1) protein is another inhibitory receptor that is induced after T cell activation. It engages with its ligands, PD-L1 and PD-L2, leading to downregulation of the immune response. Apart from these negative signals, the T cell immune response may be dampened by the decreased expression of CD40L, OX40, and 4-1BB after T cell activation (Broaddus, 2022).

3.3.3. Mechanisms of Cellular Immunity

The initiation, propagation, and regulation of cellular immune response are mediated by helper T (Th) cells, which help activate cytotoxic T cells and macrophages to attack infected cells as well as stimulate B cells to secrete immunoglobulins (Bhagavan & Ha, 2011; Britannica, 2021).

3.3.3.1. Helper T cell-mediated immune responses

Fully activated Th cells proliferate to increase in number, and they secrete different types of cytokines (see Figure 3.35). Th1cell stimulates strong cell-mediated immunity in response to intracellular pathogens by activating cytotoxic T cells. It also activates bactericidal activities and enhances the anti-viral response of macrophages, other mononuclear phagocytes, and NK cells. The TCR of the Th1 cell recognizes a specific antigen-MHC II complex on the macrophage cell surface providing the first signal to activate the macrophage (see Figure 3.37). A second signal is provided by the interaction of CD40-CD40L located on the surface of the Th1 cell and macrophage, respectively. Such interaction stimulates the Th1 cell to produce IFNγ, IFNα, TNF-β, and IL2 as well as express CXCR3 and CD161. Moreover, the activated macrophages have an enhanced ability to uptake and destroy intracellular pathogens and tumor cells. This is accomplished by enhanced expression of MHC II on their cell surface leading to increased antigen presentation, increased surface expression of co-stimulatory molecules, and improved secretion of inflammatory mediators, such as TNFα, IL1, IL12, reactive oxygen intermediates (ROI), and nitric oxide (NO). Activated macrophages are also vital in inducing and activating CD8 cytotoxic T cells. Th1 cell-derived cytokines are also involved in inducing B cells to produce opsonizing antibodies, like IgG, that enhance the efficiency of phagocytes (Cano & Lopera, 2013; Britannica, 2021; Marshall *et al.*, 2018; Bonilla & Oettgen, 2010; Janeway *et al.*, 2001; Bhagavan & Ha, 2011). Th1 cells are also associated with organ-specific autoimmunity (Luckheeram *et al.*, 2012).

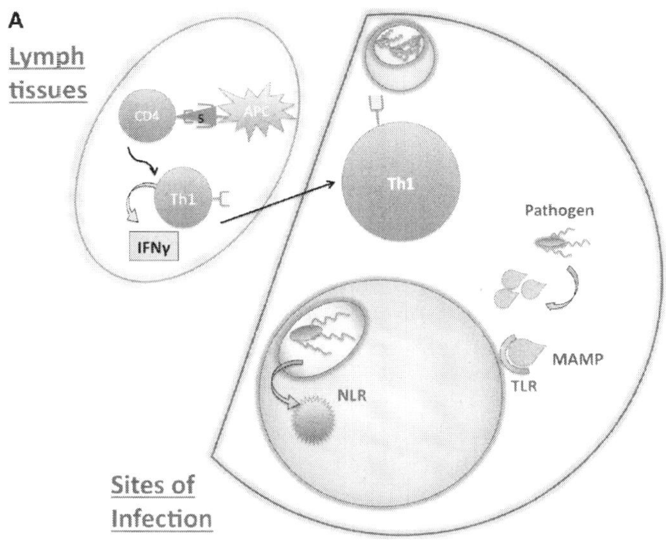

Figure 3.37. In lymph tissues, CD4 T cells are initially activated upon recognition of peptides presented by antigen presenting cells (APCs) and are further activated by a second signal. During intracellular infections, the cytokines IFNγ and IL12 trigger the differentiation of the activated CD4 T cells into helper T

cells 1 (Th1), which produce IFNγ. Then the Th1 cells move to the infection site where they are further stimulated to produce effector cytokines, which ultimately results in the destruction of the infected cell. (Source: Source: Hope O'Donnell & Stephen J. McSorley, Creative Commons License)

The cytokines secreted by Th2 cells include IL4, IL5, IL6, IL9, IL10, IL13, IL25, and amphiregulin. They also express CC chemokine receptor 4 (CCR4) and ICOS. Th2 cells initiate the humoral immune response by activating naïve antigen-specific B cells to produce IgM. Moreover, they induce immunoglobulin class switching to IgE and the development and recruitment of mast cells and eosinophils, which are essential for mounting an anti-parasitic response. Th2 cells also enhance the production of certain subtypes of IgG that function to combat bacterial infection. They are also associated with allergic reactions, especially asthma (Cano & Lopera, 2013; Britannica, 2021; Marshall *et al.*, 2018; Bonilla & Oettgen, 2010; Janeway *et al.*, 2001).

Th9 cells secrete IL9 and IL10. IL9 is a potent mast cell growth factor and mediator of helminthic immunity. Th17 cells produce IL17A, IL17F, IL9, IL21, IL22, IL25 (formerly called IL17E), IL26, and CCL20. IL17A and IL17F are potent pro-inflammatory cytokines that induce the production of IL6 and TNF and stimulate granulocyte recruitment. They mount immune response versus extracellular bacteria and fungi. Th17 cells are also associated with autoimmunity, and they have a key role in chronic allergy inflammatory processes, such as asthma. They usually occur in the pulmonary and digestive mucosa. Th22 cells release IL22 and TNFα as well as express fibroblast growth factor (FGF), IL13, and chemokines involved in angiogenesis and fibrosis. These cells also express CCR4, CCR6, and CCR10 enabling them to infiltrate the epidermis in individuals with inflammatory skin disorders (Cano & Lopera, 2013; Bonilla & Oettgen, 2010; Golubovskaya & Wu, 2016).

Treg cells produce mainly IL10, TGF-β, and IL35. They generally recognize target cells with peptide-MHC I complexes. After the pathogens have been cleared, they tone down the immune response by inhibiting the proliferation and effector function of various immune cells, such as B cells, CD4 or CD8 T cells, NK cells, NK T cells, and APC. Thus, they are critical in preventing autoimmune diseases and avoiding prolonged allergies. They play an important role in the tolerance of allogenic transplants and the fetus during pregnancy. They may also suppress antitumoral responses, which favors tumor development. Tfh cells, on the other hand, express CXCR5 but not CCR7. This enables them to relocate from the T cell zone to the B cell follicles where they induce the formation of germinal centers. Specifically, they interact with antigen-primed B cells and facilitate their conversion into plasma cells that can produce different immunoglobulin isotypes. In general, a Tfh cell produces IL21. There are Tfh cell subtypes that produce certain cytokines. Tfh1 cell secretes IFNγ that promotes IgG2a production, Tfh2 cell produces IL4 that favors IgG1 and IgE production, and Tfh10 cell releases IL10 that induces IgA secretion. Tfh cells are also involved in the formation of memory B cells (Cano & Lopera, 2013; Bonilla & Oettgen, 2010; Luckheeram *et al.*, 2012; Golubovskaya & Wu, 2016).

3.3.3.2. Cytotoxic immune response

There are CD4 T cells that have been observed to have cytotoxic activity (CD4 CTL). Most of these developed from Th1 cells, and their differentiation is also regulated by the transcription factor T-bet. T-bet induces the expression of granzyme B and perforin so that the CD4 CTLs secrete cytotoxic granules containing these molecules. They recognize peptides, whether self or non-self, presented by MHC II and then eventually kill the presenting cell, which is usually a B cell, macrophage, or DC but can also be an epithelial cell that expresses MHC II after infection (Takeuchi & Saito, 2017).

Nevertheless, the main effector of killing infected cells is the CD8 T cell. The CD8 cytotoxic T cell recognizes APCs expressing specific peptide-MHC I complexes in the central lymphoid organs. Upon activation, it upregulates the expression of the inflammatory cytokine receptor, CXCR3, allowing it to enter peripheral sites of infection. CD4 T cells also produce cytokines that enhance the recruitment of CD8 T cells into the infected site where they come in contact with more target cells. For instance, when the vaginal epithelium is infected with herpes simplex virus 2 (HSV-2), CD4 T cells enter the site of infection then produce IFNγ. This stimulates the epithelial cells to release the chemokines CXCL9 and CXCL10, which recruit CD8 T cells into the infected site through a process mediated by CXC3. Moreover, CD4 T cells activate APCs to induce a primary immune response by CD8 T cells against non-inflammatory antigens and certain viral infections (Zhang & Bevan, 2011; Nakanishi *et al.*, 2009).

Then, the CD8 T cell induces the apoptosis of its target cell via two mechanisms. The most important one uses the granzyme-perforin pathway which is a contact-dependent mechanism (see Figure 3.38). This is activated when the TCR and CD8 of a circulating cytotoxic T cell recognize an antigen-MHC I complex on the surface of a virus-infected cell or tumor cell. This leads to the formation of an immunologic synapse producing a signal that triggers the mobilization of the cytolytic granules in T cells. The granules, which are surrounded by a lipid bilayer, then fuse with the plasma membrane of the CD8 T cell and then release their contents into the immunologic synapse. The granules contain membrane pore forming proteins, serine proteases, and calreticulin. Calreticulin is a perforin inhibitor that protects the cytotoxic T cell from autolysis. There are two membrane-forming proteins in the granules, namely perforin or cytolysin and granulysin, with the latter known to be involved in degrading membrane lipids. The proteases are also known as granzymes or fragmentins, and these are complexed with a proteoglycan matrix protein known as serglycin (SG). Perforins form pores or channels in the membrane of the target cell allowing the granzymes to enter the infected or malignant cells (Bhagavan & Ha, 2011; Bonilla & Oettgen, 2010; Cano & Lopera, 2013; Marshall *et al.*, 2018). Once inside the cytosol, certain granzymes cleave the BH3-interacting domain death agonist (BID) and pro-caspase 3. The truncated BID alters the mitochondrial membrane triggering the release of pro-apoptotic factors like cytochrome C, which is involved in apoptosome formation, and endonuclease G, which fragments DNA. The activated caspase 3, on the other hand, induces the activation of endonuclease and protease, which digest DNA and cytoskeletal proteins, respectively (Perdomo-Celis *et al.*, 2019). The cytolytic granules are released only in the direction of the target cell to avoid damaging the adjacent, healthy cells. After releasing such granules, the CD8 T cells can move to a new target cell and induce its apoptosis (Wissinger, 2021).

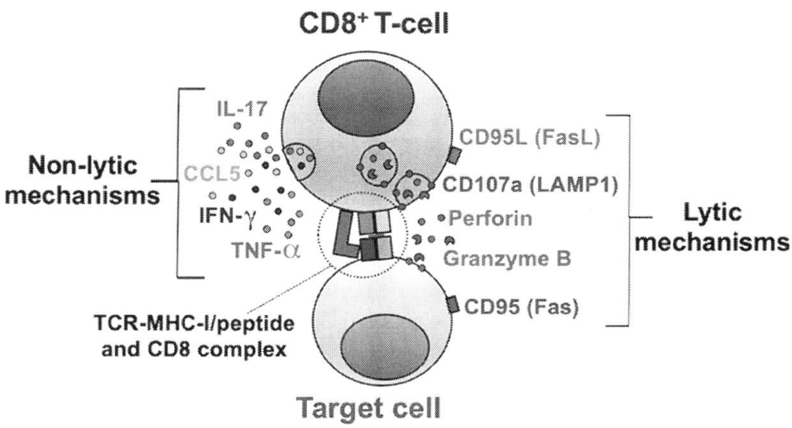

Figure 3.38. CD8 T cells can induce cytolysis by two mechanisms. One is by releasing granules containing perforin that creates pores in the target cell membrane and granzyme B that induces DNA and protein fragmentation. Another

way is via the Fas-FasL interaction that initiates a cascade that activates caspase enzymes, which mediate apoptosis. CD8 T cells also secrete cytokines, such as IFNγ, TFNα, and IL17, and the chemokine CCL5 that regulate antiviral and inflammatory processes. (Source: Federico Perdomo-Celis *et al.*, Creative Commons License)

Another mechanism employed by a CD8 cytotoxic T cell is through Fas-FasL interactions (see Figure 3.38). FasL (CD95L) is stored in specialized secretory lysosomes but upon the activation of CD8 T cells, it is expressed on the cell surface, where it binds the death receptor Fas (or CD95) that is found on most cell surfaces. The binding of FasL causes Fas to aggregate and form a trimer, which then recruits signaling molecules. Such molecules induce the activation of caspase enzymes resulting in the fragmentation of the DNA and the degradation of cytoskeletal proteins ultimately causing the death of the target cell. Since CD8 T cells can express both FasL and Fas, this mechanism can result in CD8 T cells killing each other, a process called fratricide, during the end of an immune response (Bhagavan & Ha, 2011; Wissinger, 2021). Most of the CD8 T cells die upon resolution of infection, but some cells remain as memory cytotoxic T cells that can quickly differentiate into effector cells upon subsequent encounters with the same antigen.

Cytotoxic T cells also release cytokines including IFNγ, TNFα, and TNF-β which are essential in the immune response to viral infections, inflammatory responses, and the regulation of tumor cell proliferation (see Figure 3.38) (Cano & Lopera, 2013; Marshall *et al.*, 2018). They also produce IL2 that promotes T cell survival and proliferation and IL17 that activates mucosal immunity by inducing epithelial cells to secrete defensin, AMPs, and regenerating proteins (Perdomo-Celis *et al.*, 2019). Moreover, CD8 T cells secrete many chemokines that promote the mobilization of other cells to the infected site.

CD8 T cells also have a noncytotoxic, non-MHC dependent activity called the CD8 T cell noncytotoxic antiviral response (CNAR). This has been observed in the antiviral response against all strains of human immunodeficiency virus 1 (HIV-1) and HIV-2. This is partly mediated by the CD8 T cell antiviral factor (CAF) that is secreted by CD8 T cells. CAF acts on the long terminal repeats (LTRs) of the HIV genome so that transcription of the viral genes is inhibited, which leads to lower levels of the viral proteins. As a result, there is decreased HIV replication in CD4 T cells and in macrophages (Levy & Castelli, 2019).

CD8 T cells also perform a regulatory role by secreting the anti-inflammatory cytokine IL10. This controls local inflammation and prevents excessive tissue damage at the periphery of the infected site. This IL-10 producing ability by the CD8 T cells is only temporary and it is activated by antigen exposure, IL2 from CD4 T cells, and IL27 (Zhang & Bevan, 2011).

3.4 THE MEMORY IMMUNE CELLS

The immunity mounted upon first exposure to the antigen is called the primary immune response. It is relatively slow taking a few weeks to fully develop. Memory B and T cells and NK cells are generated during the primary response. The secondary immune response is established more quickly since the same antigen induces the activation of the memory cells at a shorter amount of time, and thus prevents the associated disease from taking hold. In some cases, this occurs even after several years since the initial antigen exposure (Bonilla & Oettgen, 2010; Ratajczak *et al*., 2018).

3.4.1. Memory B Cells

Memory B cells have the same size and morphology as naïve B cells, but they express a different set of surface markers and have a longer life span with increased sensitivity to a low antigen dose. After the primary immune response, memory B cells usually reside where the antigen might be encountered again. For instance, when the antigen is first encountered in a lymph node, some memory B cells remain in that part. However, other memory B cells may leave that lymph node and enter the circulation so that they can monitor the other lymph nodes for the antigen. Memory B cells that are formed in response to antigens in the spleen stay in the splenic marginal zones, where blood-borne antigens gather (Mak *et al*., 2014). Most of the memory B cells may be found in the spleen. In fact, about 45% of the total B cell population in this organ are memory B cells (Hauser & Höpken, 2015). Some memory B cells occur in the epithelial cells of the skin and mucosal surfaces as part of skin-associated lymphoid tissues (SALT) and MALT, respectively (Mak *et al*., 2014).

Memory B cells switch from anabolic to catabolic metabolism to reduce the requirements for growth factors. They express BCL2, a protein that prevents apoptosis. Their long life span is regulated by follicular dendritic cells (FDCs) and the T-bet transcription factor (Ratajczak *et al*., 2018).

During the primary humoral immune response, IgM molecules with low affinity are predominantly produced. Other isotypes with higher affinity appear two weeks or more later. During the secondary humoral immune response, the memory B cells can immediately home to the primary follicles of a lymph node due to their increased number of adhesion molecules. Also, the immunoglobulins secreted during the secondary immune response have changed isotypes, i.e., IgG, IgA or IgE, with high affinity BCRs specific to the antigen (Bonilla & Oettgen, 2010; Ratajczak *et al*., 2018). As a result, memory B cells are activated more easily and efficiently, and they have the potential to rapidly differentiate into plasma cells so that they can quickly produce large amounts of immunoglobulins. The presence of many memory B cells can also accelerate the mounting of cellular immunity since they can serve as APCs for memory Th cells without having to wait for a DC (Mak *et al*., 2014; Ratajczak *et al*., 2018).

3.4.1.1. Types of memory B cell

There are many types of memory B cells formed. In the blood and bone marrow, the classic memory B cell produces IgM as well as expresses the surface markers CD19, CD27, and CD80. These originate from germinal centers and are able to switch immunoglobulin isotype as well as to differentiate into long-lived plasma cells. They are vital in providing immunity against bacteria producing polysaccharide capsules and merozoite surface protein 1 (MSP1). Another type has a similar phenotypic profile but also produces IgD. These resemble B cells in the mantle zone of the spleen. Other memory B cells with CD27 and CD19 produce IgA or IgG, which have IgG1, IgG2, and IgG3 subsets. These are found in the tonsil and spleen (Palm & Henry, 2019; Ratajczak *et al*., 2018).

On the other hand, there are IgM-producing memory B cells that are devoid of CD27 and express the Fc receptor-like-4 (FCRL4) protein. These are tissue residents, and they are able to rapidly differentiate into plasma cells. Then there are IgG-producing memory B cells but without CD27 surface marker. They can produce IgG subclasses of IgG1 and IgG3. Finally, there are some memory B cells that produce IgE (Palm & Henry, 2019; Ratajczak *et al*., 2018).

3.4.1.2. Generation of memory B cells

T cell-dependent antigen activation drives the interaction of B and T cells in the boundary zone or the border between B and Tcell zones of secondary lymphoid organs (see Figure 3.39a). Then the activated B cells proliferate in the outer follicle and follow one of three possible fates (see Figure 3.39b).

Some of them differentiate into short-lived, high-affinity plasma cells that secrete immunoglobulins. This ensures that the initial batch of secreted immunoglobulins has enough affinity for the antigen to opsonize it and form immune complexes to trigger phagocytosis, to activate complement, and to be presented on FDCs that subsequently induce affinity maturation in the germinal center. Others develop into low affinity memory B cells, independent of the germinal centers. These usually did not undergo somatic hypermutation or isotype switching, and they produce IgM. These may be recruited in the primary immune response or recalled during the secondary immune response. Finally, the third choice is for cells to upregulate BCL6 expression, move deeper into the follicle, and form a germinal center (Ratajczak *et al*., 2018; Mak *et al*., 2014; Palm & Henry, 2019).

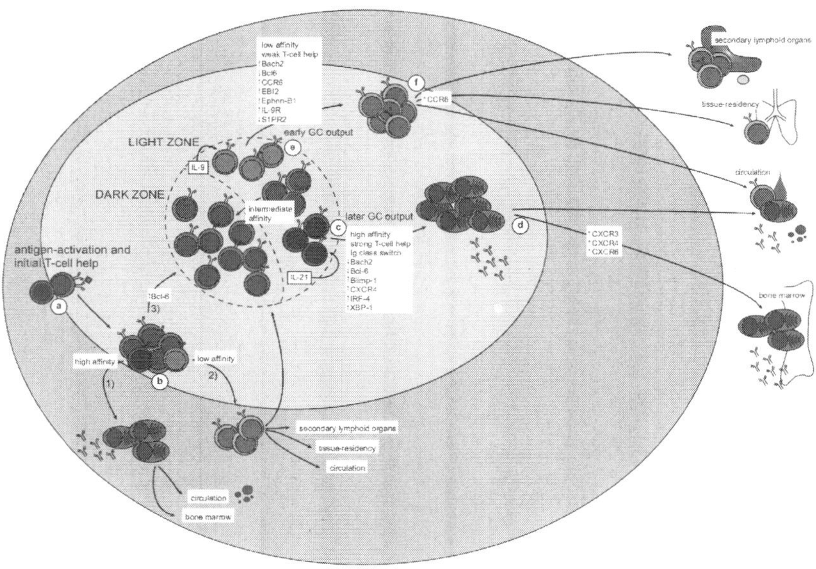

Figure 3.39. (a) In the secondary lymphoid organs, antigen activation leads to B- and T-cell interaction at the boundary of the B- and T-cell zones. (b) The activated B cells proliferate and differentiate either into (1) plasma cells, (2) very early memory B cells, or (3)cells with upregulated Bcl-6 that form a germinal center (GC). (c) In the light zone of the GC, high affinity B cells receive strong T-cell support resulting in a transcriptional profile that produces plasma cell precursors. (d) These precursors either enter circulation as short-lived plasma cells or migrate to the bone marrow, where they develop into long-lived plasma cells. (e) The low affinity B cells in the light zone receive weaker T-cell help. Thus, they will follow a different transcriptional program, which leads to memory B cell development. (f) These will then reside in secondary lymphoid

organs, become tissue residents, or enter circulation. (Source: Anna-Karin E. Palm & Carole Henry, Creative Commons License)

In the dark zone of the germinal center, antigen-specific B cells undergo clonal expansion and BCR differentiation via somatic hypermutation resulting in higher affinity to a specific antigen. These cells then migrate to the light zone, where the new BCR will be tested against the antigen presented by FDCs. Those B cells with high-affinity BCRs receive strong induction from T cells resulting in isotype switching and downregulation of Bach and B cell lymphoma 6 (BCL6) expression while upregulating Blim-1, IRF-4, XBP-1, and CXCR4 (see Figure 3.39c). This leads to the development of plasma cell precursors, which then enter circulation as short-lived plasma cells or upregulate CXCR3, CXCR4, and CXCR5 to allow migration to the bone marrow (see Figure 3.39d). In the bone marrow, stromal cells and other adjacent cells like eosinophils and macrophages release survival factors to promote the differentiation of the plasma cell precursors into long-lived plasma cells. These secrete immunoglobulins for a lifetime, and their high affinity BCRs ensure that these immunoglobulins are antigen-specific (Ratajczak *et al.*, 2018; Mak *et al.*, 2014; Palm & Henry, 2019).

Meanwhile, the low affinity B cells in the light zone of the germinal center receive weaker help from T cells and differentiate into memory B cells (see Figure 3.39e). The differentiation of memory B cells in the germinal center is controlled by signals produced by FDCs and Tfh cells, such as CD40L, SLAM-associated protein (SAP), ICOS, and BCL6 protein. The increased expression of Bach2, CCR6, EBI2, Ephrin-B1, and IL-9R also promotes the differentiation to memory cells. BCL6 is a transcriptional repressor, and Bach2 is also believed to suppress transcription. Therefore, lack of strong signaling may lead to the inhibition of the plasma differentiation pathway and forces the cell to become a memory B cell The memory B cells then locate themselves strategically in secondary lymphoid organs (i.e., marginal zone of the spleen, subcapsular sinus of lymph nodes), become residents of the infected tissues, or become recirculating cells that patrol the blood (see Figure 3.39f). This is to maximize the likelihood of encountering antigens that were confronted before, and if they do so, they provide an enhanced and quick immune response. Their low affinity BCRs also provide a high degree of flexibility in terms of antigen binding (Ratajczak *et al.*, 2018; Mak *et al.*, 2014; Palm & Henry, 2019).

Memory B cells can also be produced from marginal B2 lymphocytes in the marginal zone of the spleen white pulp. B1 (i.e., B1a and B1b) lymphocytes are also involved in memory B cell generation in the peritoneal cavity where they are most numerous and to a lesser extent, in the spleen, tonsils, and peripheral blood (Ratajczak *et al.*, 2018). Meanwhile, long-term plasma cells migrate to the bone marrow, where they can reside for decades from the time of original exposure to the antigen. They may also be found in other lymphoid organs (Palm & Henry, 2019).

3.4.1.3. Mechanism of memory B cell immunity

Re-exposure to an antigen elicits a faster, stronger, and more specific immune response due to memory recall. Initial protection is provided by circulating immunoglobulins secreted by long-lived plasma cells (see Figure 3.40a). When these are not enough to neutralize and eliminate the pathogens, then memory B cells are recalled (Palm & Henry, 2019).

As previously mentioned, memory B cells are located at strategic sites where they are most likely to encounter antigen (see Figure 3.40b). If the infected tissue has resident memory B cells, these, along with circulating memory B cells, are quickly activated to proliferate and differentiate into plasma cells. Other memory B cells reside in the splenic marginal zone and the lymph node subcapsular sinus (SCS). In these locations, they interact with macrophages expressing the CD169 surface marker which present the antigens. The memory B cells, in turn, interact with Tfh cells (Merlo & Mandik-Nayak, 2013; Palm & Henry, 2019). Upon antigen activation, the memory B cells may follow three possible fates, namely immediately form SCS proliferative foci, form new germinal centers, or directly differentiate into plasma cells. The proliferative foci produce short-lived plasma cells and new memory B cells as well as B cells that may enter the germinal center (see Figure 3.40c). The activated memory B cells that re-entered the germinal center undergo further affinity maturation and class switch recombination (CSR). The differentiation fate decided by the memory B cells depends on their BCR affinity, isotype, and differential expression of CD80 and PD-L2. All of the high affinity, $CD80^{+}$ PD-$L2^{+}$ memory B cells that underwent CSRdifferentiate directly into plasma cells (see Figure 3.40f). Also, most of the unswitched (i.e., IgM-producing) and switched (i.e., IgG- or IgA-producing) memory B cells that express one of these two markers develop into plasma cells but a few of them have retained the ability to form germinal centers (see Figure 3.40e). The germinal center is usually formed by low affinity, unswitched $CD80^{-}$ PD-$L2^{-}$memory B cells, but some of these also become plasma cells (see Figure 3.40d) (Palm & Henry, 2019).

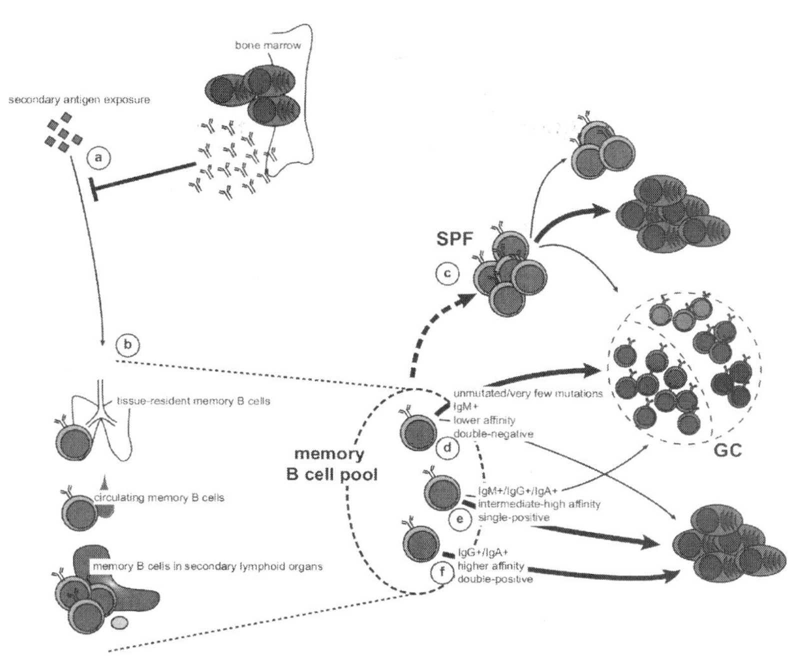

Figure 3.40. (a) The antibodies produced by long-lived plasma cells (purple) provide the first line of defense against secondary antigen exposure. (b) If the antigen persists, then tissue-resident and circulating memory B cells (green) are directly activated. Some of the antigens are carried to secondary lymphoid organs. The antigen activates memory B cells in lymph nodes to form subcapsular proliferative foci (SPF), a germinal center (GC), or plasma cells (purple). (c) SPF produce plasma cells, new memory B cells, and B cells that enter the germinal center (GC). (d) GCs are usually formed by unmutated, low affinity, IgM^+, $CD80^-$ $PD\text{-}L2^-$memory B cells, but some of these may become plasma cells. (e) Most of the IgM^+ and switched memory B cells that express either CD80 or PD-L2 directly develop into plasma cells, although they may also form GCs. (f) Switched, high affinity memory B cells that express both markers exclusively form new plasma cells. (Source: Anna-Karin E. Palm & Carole Henry, Creative Commons License)

3.4.2. Memory T Cells

Memory T cells are antigen-specific and long-lived. Compared to naïve T cells, they are more rapidly converted into effector T cells upon encountering again a peptidic antigen that is presented by MHC. They then quickly proliferate and exert their effector functions to provide a rapid immune response.

3.4.2.1. Types of memory T cell

When T cells are exposed to antigens and activated, they exhibit certain changes in terms of the surface markers L-selectin, CD44, and CD45. L-selectin is lost and CD44 levels increase. Naïve T cells express CD45RA and loses expression of CD45RO. This is markedly different from memory T cells whose CD45 shift from the high molecular weight isoform, CD45RA, to the low molecular weight type, CD45RO. Another marker, the chemokine receptor CCR7, distinguishes subpopulations of memory T cells. CCR7 mediates homing of memory T cells to lymph nodes through high endothelial venules(HEVs) (Gattorno & Martini, 2005). There are different types of memory T cells thatare in charge of pathogen surveillance of various parts of the body (see Figure 3.41).

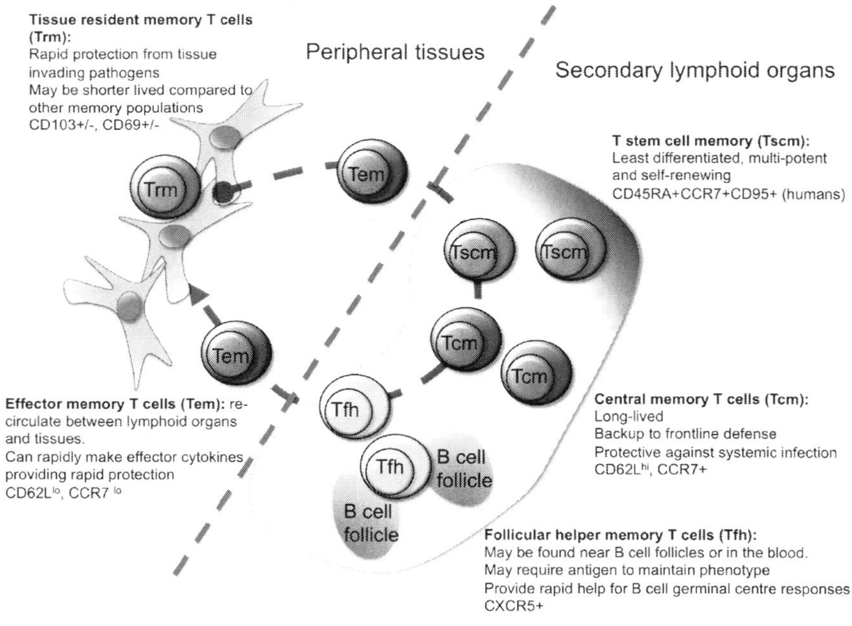

Figure 3.41. There are various types of memory T cells. Stem cell memory (Tscm) cells and central memory T (Tcm) cells are found in lymphoid organs and in the blood. They are relatively undifferentiated and long-lived. Effector memory T (Tem) cells and tissue-resident memory T (Trm) cells both occur in peripheral tissues, but Tem cells migrate between tissues and the blood, while Trm cells are restricted in tissues. They are more differentiated than Tscm and Tcm cells. Follicular helper T (Tfh) cells are effector CD4 T cells positioned near B cell follicles to assist in germinal center formation. (Source: Shafqat Ahrar Jaigirdar & Megan K.L. MacLeod, Creative Commons License)

Central memory T (Tcm) cells are found in secondary lymphoid organs, usually in lymph nodes and tonsils (see Figure 3.41). They express CD44, CD45RO, CD27, CD28, CD95, CD122, and LFA-1. They also express CD62L (or L-selectin) and CCR7, which allow their homing to secondary lymphoid organs (see Figure 3.41.and Figure 3.42). They rapidly proliferate upon reactivation, but they produce only small amounts of the cytokines IL2, IFNγ, and TNF as well as CD40L ligand. These are similar to CD8 CD122 Treg cells, but they lack PD1 protein, which confers immunosuppressive function to the Treg cells (Gattorno & Martini, 2005; Ratajczak *et al.*, 2018).

Effector memory (Tem) cells may develop from Th1 and CD8 T cells. They occur mostly in non-lymphoid peripheral tissues, such as in the lungs, liver, and intestines (see Figure 3.41). They can also recirculate between the blood and tissues. They express CD44, CD45RO, CD95, CD122, and killer cell lectin-like receptor G1 (KLRG1) protein. They do not express or express low levels of CCR7 and CD62L, but they express high levels of integrins (see Figure 3.41and Figure 3.42). This phenotype enables their migration and homing to peripheral tissues. The absence of CCR7 is also associated with their ability to secrete high levels of pro-inflammatory cytokines like IL4, IL5, and IFNγ. They also form granules with perforin and granzyme B, which confer them with cytotoxic function (Gattorno & Martini, 2005; Ratajczak *et al.*, 2018). Thus, they show effector responses faster than Tcm, but they proliferate less (Jaigirdar & MacLeod, 2015).

Some memory T cells are the tissue-resident (Trm) type. They do not circulate but instead occupy certain tissues that were previously infected. CD8 Trm cells occur in the digestive tract, female reproductive organs, lungs, skin, and the brain. As for CD4 Trm cells, they are found in the dermis, lungs, genital mucosa, intestines, and salivary glands. They have also been observed to occur in the lungs, skin, and mucosal surfaces. The Trm cells express the surface markers CD44, CD69 and CD103 (or αE). However, CD103 is not usually expressed by CD4 Trm cells in mucosal and lymphoid sites. This integrin molecule binds with β7 to form αEβ7. The αEβ7integrin molecule has E-cadherin as its ligand and along with CD49a, it enables Trm cells to enter tissues. Meanwhile, CD69 promotes tissue retention. Small quantities of CCR7 and sphingosine-1-phosphate receptor 1 (S1PR1) are expressed on the Trm cell surface. These ensure colonization of Trm cells in their location. Finally, they have low levels of CD62Lallowing their homing to peripheral tissues. The CD8 Trm cells located in the brain, skin, and intestines have increased effector function since they can synthesize IFNγ, which mediates target cell lysis, at a faster rate. Studies conducted on mouse models demonstrated that CD4 Trm cells also produce IFNγ as well as IL17(Ratajczak *et al.*, 2018; Turner & Farber, 2014).

Some memory cells have regulatory functions so that these are called memory regulatory T (mTreg) cells. Their surface marker phenotype is similar to that of Tem cells. For example, mTreg cells express CD27, CD44, CD127, CD45RO and CCR6 and exhibit low levels of CD62L. Those in the intestinal mucosa express low levels of CD44 and CD45RB. These mTreg cells, just like Tem cells, can quickly migrate to non-lymphoid tissues, such as the liver and the lungs, so that they prevent these from being damaged during secondary infection. During secondary viral infection, they produce great amounts of anti-inflammatory cytokines like IL10 that control the activity of CD4 T cells (Ratajczak *et al.*, 2018).

Memory T cells with stem-like properties (Tscm) are called such because they are like stem cells with the potential to regenerate. They have a phenotype intermediate of naïve T cells and Tem or Tcm cells. Just like naïve T cells, they express the surface markers CD27, CD45RA, and CCR7, and they do not express CD45RO (see Figure 3.41). On the other hand, they have lower expression of CD38 and CD31 compared to naïve T cells. Similar to other types of memory T cells, they show increased expression of CD95 and CD122 (IL-2Rβ). They also greatly express the antiapoptotic molecule B-cell lymphoma 2 (BCL2), CD62L, SCA1, CCR3, and CCR4. On the other hand, they express low levels or do not expressCD44, and Fas. When they express Fas and it interacts with FasL, apoptosis is induced. On the other hand, when Fas interacts with the T-cell factor (TCF) which is a transcription factor, the β-catenin signal pathway is activated. This pathway is involved in the self-renewal of cells. The unlimited proliferation of Tscm cells is facilitated by IL7 and IL15. They also secrete IL2 and IFNγ (Ratajczak *et al.*, 2018; Flynn & Gorry, 2014; Jaigirdar & MacLeod, 2015).

3.4.2.2 Generation of memory T cells

The developmental pathway for a T cell to become a memory cell is not yet clearly elucidated so that various explanations are forwarded. The linear model provides a pathway in which the activated T cells undergo expansion and become a homogeneous population of effector T cells. During the contraction phase, however, 90-95% of the T cells undergo apoptosis and the remaining 5-10% turn into memory T cells, first into Tem then into Tcm cells (see Figure 3.42a). Another model, the asymmetrical or bifurcative one, proposes that the decision to become a memory cell is made very early in the immune response during the first asymmetric cell division, wherein one daughter cell becomes a short-lived effector cell while the other is committed to follow the memory cell pathway (see Figure 3.42b) (Ratajczak *et al*., 2018; Jameson & Masopust, 2018). The self-renewal model proposes that naïve T cells develop initially into self-renewing Tcm or effector T cells. These further develops into Tem cells, which become terminally-differentiated effector cells in the non-lymphoid tissues (see Figure 3.42c). Finally, the simultaneous model suggests that the type of CD4 T cell will determine the memory T cell type that will be generated. For instance, Th1 or Th17 cells may differentiate into Tem cells, while Tfh cell, with the help of B cells, gives rise to Tcm cells (see Figure 3.42d) (Raphael *et al*., 2020).

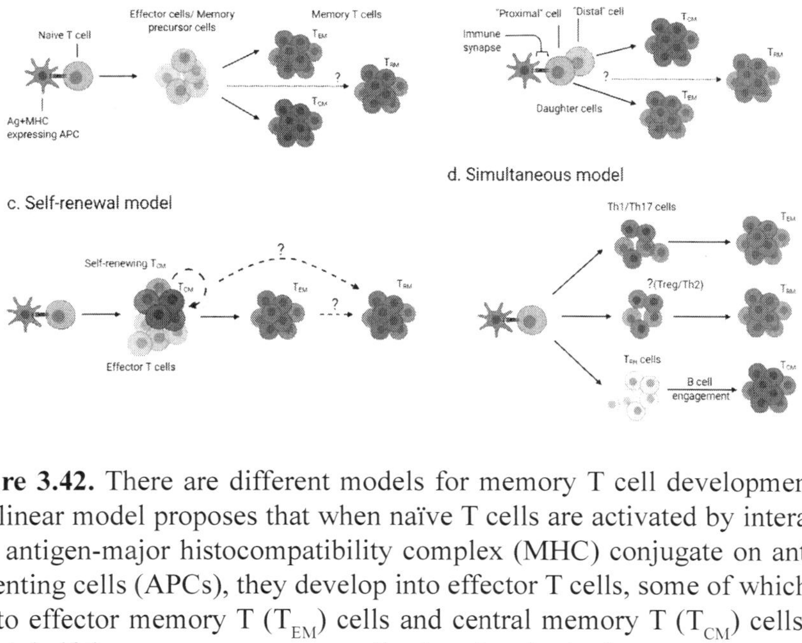

Figure 3.42. There are different models for memory T cell development. (a) The linear model proposes that when naïve T cells are activated by interacting with antigen-major histocompatibility complex (MHC) conjugate on antigen-presenting cells (APCs), they develop into effector T cells, some of which give rise to effector memory T (T_{EM}) cells and central memory T (T_{CM}) cells. It is uncertain if the precursor memory cells also give rise to tissue-resident memory T (T_{RM}) cells. (b) In the asymmetrical model, the daughter cells proximal to the immune-synapse develop into T_{EM} cells, while those distal daughter cells become T_{CM} cells. The origin of T_{RM} cells is unknown. (c) In the self-renewal model, naïve T cells give rise to self-renewing effector T cells or T_{CM} cells, which give rise to T_{EM} cells. T_{RM} cells may possibly come from the self-renewing cells or from the T_{EM} cells. (d) The simultaneous model proposes that naïve T cells differentiate into different types of T cells, which then differentiate into different types of memory cells. Helper T cell 1 (Th1) and Th17 become T_{EM}cell, while follicular helper T (T_{FH}) cell gives rise to T_{CM} cell. The Th cell type that

differentiates into T_{RM} is still unknown. (Source: Itay Raphael *et al.*, Creative Commons License)

The asymmetrical model is supported by some data showing that CD8 T cells that were produced after the first division showed "pre-effector" or "pre-memory" gene expression profiles. However, further studies indicate that pre-effector daughter cells do not exclusively differentiate into short-lived effector cells since they can also develop into long-lived effector cells. These long-lived cells have a phenotype that is intermediate to short-lived effector cells and Tem cells. On the other hand, pre-memory cells usually produce Tcm and Tem. Thus, the initial asymmetric division may be less likely to predestine whether a daughter cell persists to memory phase or not, but rather produces daughter cells that are likely to become various types of memory cells (Jameson & Masopust, 2018). CD4 IFNγ-producing T cells do not efficiently develop into resting memory cells. However, CD4 T cells activated in Th1-polarizing conditions that have not yet become IFNγ-producing cells can differentiate into long-term memory cells. Th2 cells have also been observed to develop into memory T cells (Seder & Ahmed, 2003). These memory cells may be the Tcm or the Tem subtype (see Figure 3.43). On the other hand, Treg cells may differentiate into Tcm or Treg effector memory cells (Golubovskaya & Wu, 2016).

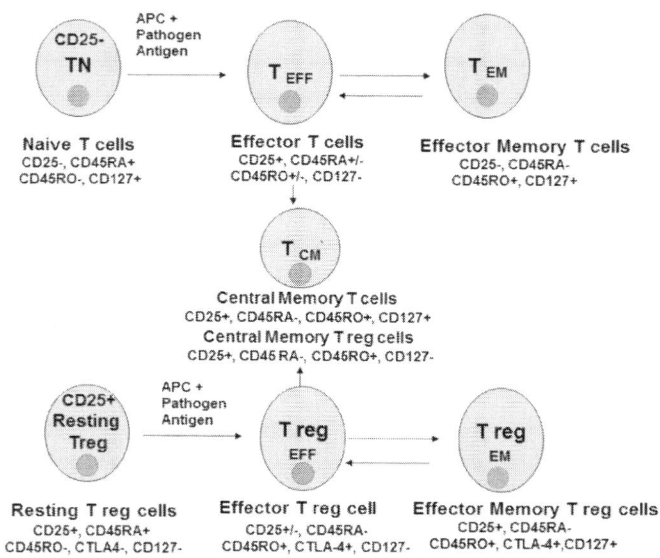

Figure 3.43. CD4 effector T cells (T_{EFF}) or regulatory cells ($Treg_{EFF}$) may differentiate into central memory T cells (T_{CM}). T_{EFF} may also differentiate into

effector memory T (T_{EM}) cells, whereas Treg cells may develop into Treg effector memory (Treg $_{EM}$) cells. During differentiation, they acquire or lose the expression of certain cell surface markers. (Source: Vita Golubovskaya & Lijun Wu, Creative Commons License)

IL2 has a vital role in CD4 memory T cell development via its regulation of BCL6 expression (see Figure 3.45). It is suggested that naïve CD4 T cells require high levels of IL2 to differentiate into memory precursor cells. These precursor cells still require IL2 signals at the later stages to develop into long-lived memory cells. IL2 drives the expression of STAT5, BCL6, and the IL7 receptor enhancing the survival and maturation of CD4 memory T cells (Raphael *et al*., 2020).

A modified theory of linear hierarchical system is proposed for CD8 memory T cell development. It proposes that naïve CD8 T cells are stimulated upon antigen exposure to differentiate first into Tscm cells (see Figure 3.44). These cells represent the earliest and longest-lasting developmental stage of memory T cells. The generation of Tscm does not appear to require continued exposure to the antigen. They are the least differentiated type that eventually turn into Tcm cells, which are long-lived memory cells. In turn, Tcm cells can transition into Tem cells wherein they progressively lose the expression of lymph node homing receptors, namely CCR7 and CD62L. They also respond to homeostatic cytokines like IL7 and IL15, tissue homing receptors, and antiapoptotic molecules. As a result, they gain the capacity for rapid proliferation after antigen re-exposure, IL2 production, and survival. It may be that the longevity of memory T cells is due to their ability to respond to IL15. These features typical of a memory cell are acquired gradually only after the clearance of the antigen. Meanwhile, the cytotoxic ability and IFNγ-producing ability of these T cells did not diminish with time (Flynn & Gorry, 2014; Seder & Ahmed, 2003; Burleson *et al*., 2015). The most differentiated state is the CD8 effector T cell (Golubovskaya & Wu, 2016). Then theCD8 effector T cells that survived apoptosis during the contraction phase become long-term resting memory cells (Seder & Ahmed, 2003). Those that express high levels of CD127, the specific receptor of IL7 that is critical for T cell survival, are more likely to develop into memory cells (Jaigirdar & MacLeod, 2015).

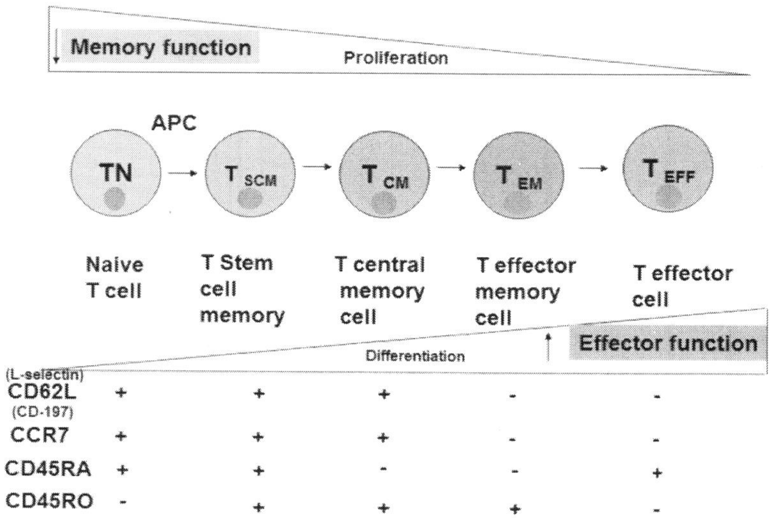

Figure 3.44. Naïve CD8 T cells may differentiate into memory stem T (T_{SCM}) cells, which in turn, can develop into central memory T (T_{CM}) cells, then into effector memory T ($_{TEM}$) cells. Each type of CD8 memory T cell expresses a distinct set of cell surface markers. Also, the potential to proliferate decreases as the T cell differentiates, whereas its effector function gets more enhanced. (Source: Vita Golubovskaya & Lijun Wu, Creative Commons License)

Whether Tem cells develop from Tcm cells or directly differentiate from Tscm cells remains unclear. Also, Tcm and Tem cell types have been observed to convert to the other, i.e., some Tcm cells become Tem cells after proliferation and expansion, and vice versa (see Figure 3.46). These two types may develop from either CD4 or CD8 T cells, but the Tcm type is more common within the CD4 subset, while the Tem cells develop more commonly from the CD8 T cells (Burleson *et al*., 2015).

The strength of TCR signaling is critical for memory T cell development (see Figure 3.45). This is determined by the affinity of the TCR to the antigen-MHC complex, the density of these complexes on APCs, and the duration of the TCR interaction with the complex. Changes in these parameters influence T cell transcriptional profiles to either promote or inhibit memory T cell formation. For CD8 memory T cells, TCR signaling strength is generally negatively correlated with memory cell formation. On the other hand, the association is less clear in the case of CD4 memory T cell generation. (Raphael *et al*., 2020).

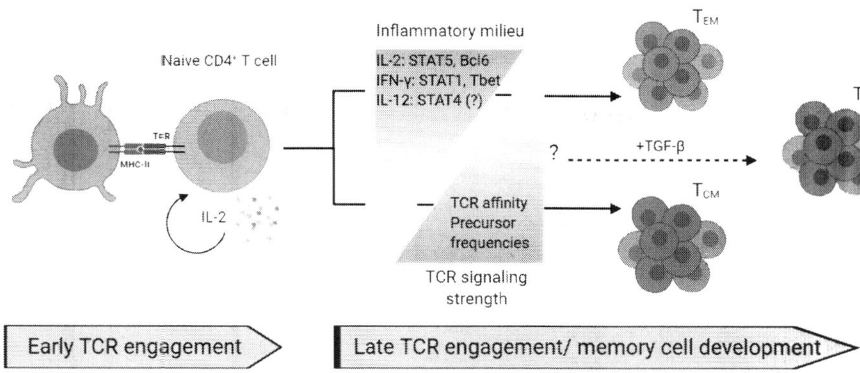

Figure 3.45. High levels of interleukin 2 (IL2) are vital for the development of naïve CD4 T cells into memory T cells. Such development is dependent on the level of inflammatory cytokines, TCR affinity, and precursor frequencies. Increased levels of inflammatory signals favor the formation of effector memory T (T_{EM}) cells, while increased TCR signaling and precursor frequencies lead to the formation of central memory T (T_{CM}) cells. TGF-β is known to be vital for tissue-resident memory (T_{RM}) cells, but the involvement of other inflammatory signals and TCR signaling remains unclear. (Source: Itay Raphael *et al.*, Creative Commons License)

Another factor that affects memory T cell formation is the frequency of naïve T cell precursors, which is defined by the fraction of naïve T cells that can respond to a specific epitope and to enter the effector or memory antigen-specific T cell pool (see Figure 3.45). Excessively high frequencies reduce the proliferation and differentiation of antigen-specific T cells upon infection, thus impairing memory T cell generation. In general, increased levels of TCR affinity and precursor frequencies favor Tcm cell development, while decreased levels favor the formation of Tem cells (Raphael *et al.*, 2020).

Since TCR signaling appears to regulate the ratios of several transcription factors, there are changes in gene expression as naïve T cells progressively become Tscm then Tcm and Tem cells (see Figure 3.46). There is a downregulation of transcription factors, such as lymphoid enhancer-binding factor 1, forkhead box P1, and LAG1 homolog, which inhibit T cell activation and differentiation from naïve to Tem cells. In contrast, there is upregulated expression of regulators of effector differentiation (e.g., eomesodermin and T-box 21), cytotoxic molecules (e.g., granzyme A and perforin), and T cell senescence (e.g., KLRG1) (Gattinoni *et al.*,

2011). KLRG1 expression is also observed in Trm cells. In addition, the expression of transcription factors EOMES, TCF1, BCL6, ID3, and STAT3 is associated with the development of Tcm cells, whereas the transcription factors T-bet, Blimp1, ID2, and STAT4 are related to the Tem phenotype (Martin & Badovinac, 2018).

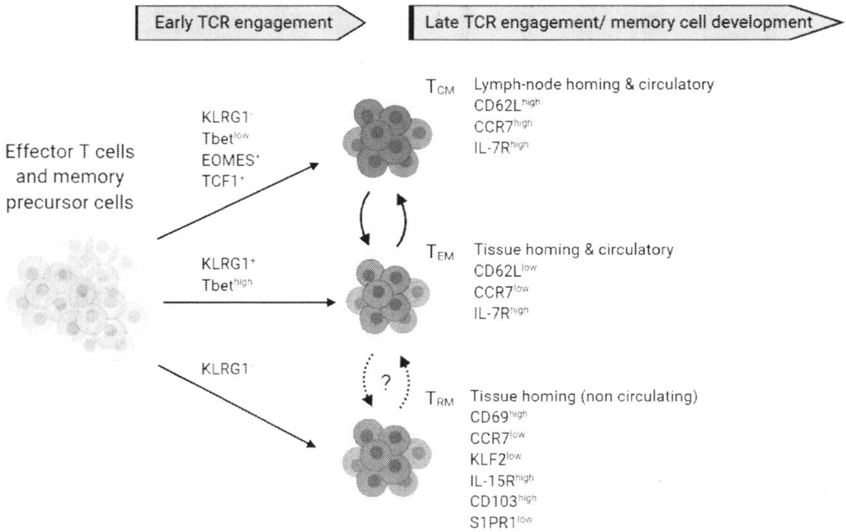

Figure 3.46. Differential expression of certain regulatory proteins is observed as T cells develop into different types of memory T cells. The expression of EOMES and TCF1 is upregulated for central memory T (T_{CM}) cells. Meanwhile, KLRG1 and Tbet are highly expressed in effector memory T (T_{EM}) cells and KLGR1 expression is observed in tissue-resident memory T (T_{RM}) cells. T_{CM} cells can give rise to T_{EM} cells and vice versa. (Source: Itay Raphael *et al.*, Creative Commons License)

Less is known about the development of Trm cells. Effector T cells from lymphoid organs that respond to infection or inflammation within non-lymphoid tissues may further respond to certain regulatory or tissue-specific factors, which enable them to develop a tissue resident phenotype (see Figure 3.47). Homing of Trm cells in specific tissues is mediated by the expression of various integrins and chemokine receptors, which facilitate T cell migration into certain tissues. For instance, chemokine receptor 9 (CCR9) and the integrin α4β7 target T cells to the intestines, cutaneous leukocyte antigen (CLA) transport cells to the skin, and lung DC upregulates CCR4 to promote migration of T cells to the lungs (Turner & Farber, 2014).

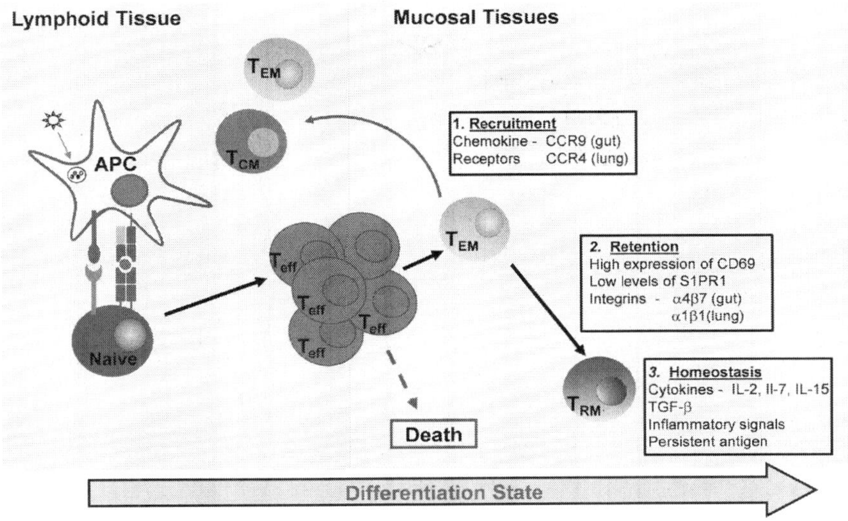

Figure 3.47. (1) Tissue resident memory T (T_{RM}) cells are likely derived from effector T cells that are recruited by certain chemokines to migrate into the target tissues. Most of the effector T cells die but some differentiate into various types of resting memory T cells including central memory T (T_{CM}) cells (migrate back to lymphoid tissues), effector memory T (T_{EM}) cells (circulate through peripheral tissues), and T_{RM} cells. (2) The T_{RM} cells are retained in the mucosal tissues possibly via the inhibition of the sphingosine-1-phosphate receptor 1 (S1PR1) that promotes lymphocyte exit from tissues and via the expression of integrins that facilitate cell-to-cell interactions. (3) The homeostasis of T_{RM} cells may depend on cytokines that maintain their survival, constitutive low-level inflammation, and, in some cases, the persistence of antigen at the site. (Source: Damian Lanz Turner & Donna L. Farber, Creative Commons License)

The maturation of CD8 Trm cells in the brain, sensory ganglia, and the lungs, requires an antigen. This is in contrast with the formation of certain CD8 Trm cells and CD4 Trm cells. Instead, CD8 Trm cells in the female reproductive system and in the intestines depend on tissue cytokines like TGFβ, TNFα, IL33, and IL15 (see Figure 3.47) (Ratajczak *et al.*, 2018). TGFβ induces the expression of the mucosal integrin, αEβ7, which enables the CD8 Trm cells to be retained in non-lymphoid tissues by promoting their interaction with local epithelial cells. This integrin is downregulated in CD4 Trm cells so that the retention of this type of Trm cells may be due to another, unknown mechanism. The lectin CD69, on the other hand, is constitutively expressed on CD4 and CD8 Trm cells. This inhibits S1PR1, which mediates the movement of lymphocytes out of tissues (Turner & Farber, 2014; Jaigirdar & MacLeod, 2015). The transcription factors Blimp1, Hobit, Nur77, and Runx3 also promote the formation of Trm cells (Jameson & Masopust, 2018).

3.4.2.3. Mechanism of memory T cell immunity

Similar to naïve T cells, Tcm cells encounter antigen exclusively in lymphoid tissues. There are more Tcm cells (i.e., as much as a thousand-fold higher quantity) than naïve ones of the same specificity. Following local viral infection, type I IFN and IFN-γ produced by activated CD8 T cells induce CXCL10 and CXCL9. This stimulates the chemokine receptor CXCR3to efficiently move activated CD4 T cells and more Tcm cells to inflammation sites in the lymph nodes, where they can encounter APCs and stimulatory cytokines(Jameson & Masopust, 2018). The IL2 produced by Tcm cells also provides them with a greater capacity to proliferate. All these provide Tcm cells with more chances of interacting with APCs presenting a previously encountered antigen so that they are immediately activated to provide a rapid and stronger immune response and to proliferate quickly generating a large pool of antigen-specific effector T cells. In addition, CD4 Tcm with Tfh cell phenotype facilitates B cell activation resulting in faster B cell expansion, quicker CSR, and increased secretion of immunoglobulins (Raphael *et al*., 2020).

The respiratory organs previously infected with SARS or MERS viruses are colonized with CD4 Tem cells. Upon re-exposure even to very low concentrations of the virus, these cells quickly secrete cytokines that activate the immune response. They also drive the migration of DCs to mediastinal lymph nodes and help direct to the lungs virus-specific CD8 T lymphocytes (Ratajczak *et al*., 2018). Tem cells have a longer lifespan than effector T cells and so they can quickly provide antigen-specific effector T cells in peripheral tissues as the need arises rather than wait for the differentiation of naïve T cells into effector T cells (Raphael *et al*., 2020).

In the case of Trm cells, they are strategically positioned as the first line of defense providing a rapid local immune response to invading pathogens. CD4 and CD8 Trm cells quickly clear viral, bacterial, and parasitic pathogens in mucosal tissues, skin, and even certain organs (Jameson & Masopust, 2018). The epidermal Trm cells have dendritic protrusions that efficiently probe their surroundings to detect the presence of pathogens. Once they are activated via their antigen-specific TCR, they initiate antimicrobial defense by inducing the expression of genes involved in antibacterial and antiviral response, such as the production of IFNγ, TNF, and IL2. These cytokines drive the activation and migration to the infected site of innate and adaptive immune cells like NK cells, DCs, B cells, and T cells (Ratajczak *et al*., 2018; Gebhardt & Bedoui, 2016). IFNγ also acts on stromal and parenchymal cells in their vicinity to confer them resistance against viruses, thereby preventing the viral pathogen from spreading further. Trm cells may also directly lyse the infected target cells (Gebhardt & Bedoui, 2016).

CHAPTER REFERENCES

1. Akbar, R., Robert, P.A., Pavlović, Jeliazkov, J.R., Snapkov, I., Slabodkin, A., Weber, C.R., Scheffer, L., Miho, E., Haff, I.H., Haug, D.T.T., Lund-Johansen, F., Safonova, Y., Sandve, G.K. & Greiff, V. (2021). A compact vocabulary of paratope-epitope interactions enables predictability of antibody-antigen binding. *Cell Reports* 34(11): 108856. https://doi.org/10.1016/j.celrep.2021.108856 (Accessed on 11 November 2021).
2. Aktor, J.K. (2012). *Elsevier's Integrated Review Immunology and Microbiology (2nd edn.)*. Pennsylvania: Elsevier Saunders. pp.17-24.
3. Alberts, B., Johnson, A., Lewis, J., Raff, M., Roberts, K. & Walter, P. (2002). *Molecular Biology of the Cell (4th edn.)*. New York: Garland Science.
4. Albrecht, M. & Arck, P.C. (2020). Vertically Transferred Immunity in Neonates: Mothers, Mechanisms and Mediators. *Frontiers in Immunology* 11:555. https://doi.org/10.3389/fimmu.2020.00555 (Accessed on 6 December 2021).
5. Allman, D. & Northrup, D.L. (2010). B-Cell Development. In: *Comprehensive Toxicology, Vol.5*; McQueen, C.A. (Ed.). Philadelphia, Pennsylvania: Elsevier Limited. pp.35-52. https://doi.org/10.1016/B978-0-08-046884-6.00603-5 (Accessed on 17 November 2021).
6. Backhaus, O. (2018). Generation of Antibody Diversity. In: *Antibody Engineering*; Thomas Böldicke (Ed.). IntechOpen. https://doi:10.5772/intechopen.72818 (Accessed on 11 October 2021).
7. Bhagavan, N.V. & Ha, C.-E. (2011). *Essentials of Medical Biochemistry: With Clinical Cases*. London: Academic Press. pp. 449-472. https://doi.org/10.1016/B978-0-12-095461-2.00033-3 (Accessed on 8 December 2021).
8. Bio-Rad (2021). *Mini-review: An overview of T cell receptors*. https://www.bio-rad-antibodies.com/t-cell-receptor-minireview.html?JSESSIONID_STERLING=073C2C6BC5BEC49C5DBD55E35042FEB5.ecommerce2&evCntryLang=PH-en&cntry=PH&thirdPartyCookieEnabled=true (Accessed on 17 December 2021).
9. Birdsall, H.H. & Casadevall, A. (2020). Adaptive Immunity: Antibodies and Immunodeficiencies. In: *Mandell, Douglas, and Bennett's Principles and Practice of Infectious Diseases (9th edn.)*; Bennett, J.E., Dolin, R. & Blaser, M.J. (Ed.). Philadelphia, Pennsylvania: Elsevier.

pp. 35-50.e2

10. Blair, P.A. & Bosma, A. (2021). Generation of B-cell/antibody diversity. *British Society for Immunology*. https://www.immunology.org/public-information/bitesized-immunology/immune-development/generation-b-cell-antibody-diversity (Accessed on 10 October 2021).
11. Boegel, S., Castle, J.C., Kodysh, J., O'Donnell, T. & Rubinsteyn, A. (2019). Cancer Immunotherapy: Bioinformatic methods for cancer neoantigen prediction. In: *Progress in Molecular Biology and Translational Science, Volume 164*; Teplow, D.B. (Ed.). Cambridge, Massachusetts: Academic Press. pp.25-60. https://doi.org/10.1016/bs.pmbts.2019.06.016 (Accessed on 16 December 2021).
12. Bonilla, F.A. & Oettgen, H.C. (2010). Adaptive immunity. *The Journal of Allergy and Clinical Immunology* 125(2): S33-40. https://doi.org/10.1016/j.jaci.2009.09.017 (Accessed on 1 November 2021).
13. Breedveld, A. & van Egmond, M. (2019). IgA and FcαRI: Pathological Roles and Therapeutic Opportunities. *Frontiers in Immunology* 10: 553. https://doi: 10.3389/fimmu.2019.00553 (Accessed 28 September 2021).
14. Britannica, 2021. *Immune system*. https://www.britannica.com/science/immune-system/T-cell-antigen-receptors (Accessed on 27 October 2021).
15. Broaddus, V.C. (2022). Adaptive Immunity. In: *Murray & Nadel's Textbook of Respiratory Medicine*; Fontenot, A.P., Atif, S.M. & Reinhardt, R.L. (Ed.). Philadelphia, Pennsylvania: Elsevier Limited. pp. 208-227.e7.
16. Burleson, G.R., Burleson, S.C.M. & Burleson, F.G. (2015). Pulmonary Immunology of Infectious Disease. In: *Comparative Biology of the Normal Lung (2nd edn.)*; Parent, R.A. (Ed.). London: Academic Press. pp. 581-600. https://doi.org/10.1016/B978-0-12-404577-4.00030-8 (Accessed on 3 January 2022).
17. Camiolo, M., Gauthier, M., Cohn, L. & Ray, A. (2020). Biology of Lymphocytes. In: *Middleton's Allerfy: Principles and Practice*; Burks, A.W., Holgate, S.T., O'Hehir, R.E., Broide, D.H., Bacharier, L.B., Hershey, G.K.K. & Peebles, R.S. (Eds.). Philadelphia, Pennsylvania: Elsevier Inc. pp. 11, 176-190.e1.
18. Cano, R.L.E. & Lopera, D.E. (2013). Introduction to T and B lymphocytes. In: *Autoimmunity: From Bench to Bedside*; Anaya, J.M.,

Shoenfeld, Y., Rojas-Villaraga, A., Levy, R.A. & Cervera, R. (Ed.). Bogota, Colombia: El Rosario University Press.

19. Chi, X., Li, Y. & Qiu, X. (2020). V(D)J recombination, somatic hypermutation and class switch recombination of immunoglobulins: mechanism and regulation. *Immunology* 160(3): 233-247. https://doi.org/10.1111/imm.13176 (Accessed on 18 October 2021).
20. de Sousa-Pereira, P. & Woof, J.M. (2019). IgA: Structure, Function, and Developability. *Antibodies* 8(4): 57.https://doi.org/10.3390/antib8040057 (Accessed on 24 September 2021).
21. de Taeye, S.W., Rispens, T. & Vidarsson, G. (2019). The Ligands for Human IgG and Their Effector Functions. *Antibodies* 8(2): 30. https://doi.org/10.3390/antib8020030 (Accessed on 23 September 2021).
22. Flynn, J.K. & Gorry, P.R. (2014). Stem memory T cells (TSCM)—their role in cancer and HIV immunotherapies. *Clinical & Translational Immunology* 3(7): e20. https://doi: 10.1038/cti.2014.16 (Accessed on 31 December 2021).
23. Gattinoni, L., Lugli, E., Ji, Y., Pos, Z., Paulos, C.M., Quigley, M.F., Almeida, J.R., Gostick, E., Yu, Z., Carpenito, C., Wang, E., Douek, D.C., Price, D.A., June, C.H., Marincola, F.M., Roederer, M. & Restifo, N.P. (2011). A human memory T-cell subset with stem cell-like properties. *Nature Medicine* 17(10): 1290-1297. https://doi: 10.1038/nm.2446 (Accessed on 31 December 2021).
24. Gattorno, M. & Martini, A. (2005). The Immune System and the Inflammatory Response. In: *Textbook of Pediatric Rheumatology (5th edn.)*; Cassidy, J.T., Petty, R.E., Laxer, R.M. & Lindsley, C.B. (Ed.). Philadelphia: Pennsylvania: Elsevier Saunders. pp.19-63. https://doi.org/10.1016/B978-1-4160-0246-8.50009-7 (Accessed on 29 December 2021).
25. Gerbhardt, T. & Bedoui, S. (2016). Recirculating and Resident Memory CD8$^+$ T Cells. In: *Encyclopedia of Immunobiology, Vol. 3*; Ratcliffe, M. (Ed.). London: Academic Press. pp.344-352. https://doi.org/10.1016/B978-0-12-374279-7.08008-5 (Accessed on 3 January 2022).
26. Golubovskaya, V. & Wu, L. (2016). Different Subsets of T Cells, Memory, Effector Functions, and CAR-T Immunotherapy. *Cancers (Basel)* 8(3): 36. https://doi: 10.3390/cancers8030036 (Accessed on 3 January 2022).
27. Gong, S. & Ruprecht, R.M. (2020). Immunoglobulin M: An Ancient

Antiviral Weapon-Rediscovered. *Frontiers in Immunology* 11 1943. https://doi: 10.3389/fimmu.2020.01943 (Accessed on 22 September 2021).

28. Hauser, A.E. & Höpken, U.E. (2015). B Cell Localization and Migration in Health and Disease. In: *Molecular Biology of B Cells (2nd edn.)*; Alt, F.W., Honjo, T., Radbruch, A. & Reth, M. (Ed.). London: Academic Press. pp. 187-214. https://doi.org/10.1016/B978-0-12-397933-9.00012-6 (Accessed on 27 December 2021).
29. Hotinger, J.A. & May, A.E. (2020). Antibodies Inhibiting the Type III Secretion System of Gram-Negative Pathogenic Bacteria. *Antibodies* 9(3): 35.https://doi.org/10.3390/antib9030035 (Accessed on 1 December 2021).
30. Jameson, S.C. & Masopust, D. (2018). Understanding Subset Diversity in T Cell Memory. *Immunity* 48.
31. Janeway, C.A., Jr., Travers, P., Walport, M. & Shlomchik, M.J. (2001). Immunobiology: The Immune System in Health and Disease (5th edn.). New York: Garland Science.https://www.ncbi.nlm.nih.gov/books/NBK27156/ (Accessed on 8 September 2021).
32. Jeurissen, A., Ceuppens, J.L. & Bossuyt, X. (2004). T lymphocyte dependence of the antibody response to 'T lymphocyte independent type 2' antigens. *Immunology* 111(1): 1-7. https://doi: 10.1111/j.1365-2567.2003.01775.x (Accessed on 24 November 2021).
33. Jones, K., Savulescu, A.F., Brombacher, F. & Hadebe, S. (2020). Immunoglobulin M in Health and Diseases: How Far Have We Come and What Next? *Frontiers in Immunology* 11: 595535. https://doi: 10.3389/fimmu.2020.595535 (Accessed on 23 September 2021).
34. Justiz Vaillant, A.A., Jamal, Z. & Ramphul, K. (2021). Immunoglobulin. Florida: StatPearls Publishing. https://www.ncbi.nlm.nih.gov/books/NBK513460/ (Accessed on 8 September 2021).
35. Keyt, B.A., Baliga, R., Sinclair, A.M., Carroll, S.F. & Peterson, M.S. (2020). Structure, Function, and Therapeutic Use of IgM Antibodies. *Antibodies* 9(4): 53. https://doi.org/10.3390/antib9040053 (Accessed on 22 September 2021).
36. Ledsgaard, L., Jenkins, T.P., Davidsen, K., Krause, K.E., Martos-Esteban, A., Engmark, M., Andersen, M.R., Lund, O. & Laustsen, A.H. (2018). Antibody Cross-Reactivity in Antivenom Research. *Toxins* 10(10): 393. https://doi.org/10.3390/toxins10100393 (Accessed on 15

November 2021).

37. Levy, J.A. & Castelli, J.C. (2019). HIV (Human Immunodeficiency Virus). In: *Encyclopedia of Cancer (Third edn.)*; Boffetta, P. & Hainaut, P. (Ed.). Philadelphia, Pennsylvania: Elsevier, Inc. pp.218-226. https://doi.org/10.1016/B978-0-12-801238-3.11085-2 (Accessed on 22 December 2021).
38. Luckheeram, R.V., Zhou, R., Verma, A.D. & Xia, B. (2012). $CD4^+$ T Cells: Differentiation and Functions. *Journal of Immunology Research* 925135.https://doi.org/10.1155/2012/925135 (Accessed on 21 December 2021).
39. Mak, T.W., Saunders, M.E. & Jett, B.D. (2014). *Primer to the Immune Response (2nd Edition)*. Burlington, Massachusetts: Academic Press Cell.
40. Marshall, J.S., Warrington, R., Watson, W. & Kim, H.L. (2018). An introduction to immunology and immunopathology. *Allergy, Asthma & Clinical Immunology* 14: 49. https://doi.org/10.1186/s13223-018-0278-1 (Accessed on 31 October 2021).
41. Martin, M.D. & Badovinac, V.P. (2018). Defining Memory CD8 T Cell. *Frontiers in Immunology* 9: 2692. https://doi: 10.3389/fimmu.2018.02692 (Accessed on 4 January 2021).
42. Merck KGaA (2021). *An Introduction to Antibodies: Antigens, Epitopes and Antibodies*. https://www.sigmaaldrich.com/PH/en/technical-documents/technical-article/protein-biology/elisa/antigens-epitopes-antibodies?gclid=CjwKCAjwn8SLBhAyEiwAHNTJbRzykFSqHa8Nu8EWf6QMVjwQO6In6afdIPHw75laWW-LNmJ9pn-HJRoCkYIQAvD_BwE (Accessed on 26 October 2021).
43. Merlo, L.M.F. & Mandik-Nayak, L. (2013). Adaptive Immunity: B Cells and Antibodies. In: *Cancer Immunotherapy: Immune Suppression and Tumor Growth (Second Edition)*; Prendergast, G.C. & Jaffee, E.M. (Ed.). San Diego, California: Academic Press. pp. 25-40. https://doi.org/10.1016/B978-0-12-394296-8.00003-8 (Accessed on 27 December 2021).
44. Mingueneau, M. (2016). Functional Cell Biology. In: *Encyclopedia of Cell Biology Vol. 3.*; Bradshaw, R.A. & Stahl, P.D. (Eds.). Cambridge, Massachusetts: Academic Press. pp.528-540. https://doi.org/10.1016/B978-0-12-394447-4.30067-0 (Accessed on 16 December 2021).
45. Mirzaei, H.R. (2020). Adaptive Immunity. In: *Reference Module in*

Biomedical Sciences. https://doi.org/10.1016/B978-0-12-818731-9.00028-8 (Accessed on 8 September 2021).

46. Moticka, E.J. (2016). Hallmarks of the Adaptive Immune Responses. In: *A Historical Perspective on Evidence-Based Immunology*. Amsterdam: Elsevier, Inc.. pp. 9-19. https://doi.org/10.1016/B978-0-12-398381-7.00002-2 (Accessed on 8 September 2021).
47. Motran, C.C., Silvane, L., Chiapello, L.S., Theumer, M.G., Ambrosio, L.F., Volpini, X., Celias, D.P. & Cervi, L. (2018). Helminth Infections: Recognition and Modulation of the Immune Response by Innate Immune Cells. *Frontiers in Immunology* 9: 664. https://doi: 10.3389/fimmu.2018.00664 (Accessed on 2 December 2021).
48. Nakanishi, Y., Lu, B., Gerard, C. & Iwasaki, A. (2009). CD8$^+$ T lymphocyte mobilization to virus-infected tissue requires CD4$^+$ T-cell help. *Nature* 462: 510-513. https://doi.org/10.1038/nature08511 (Accessed on 23 December 2021).
49. Neurath, A.R. (2008). Immune Response to Viruses: Antibody-Mediated Immunity. *Encyclopedia of Virology* 2008: 56-70. https://doi: 10.1016/B978-012374410-4.00591-4 (Accessed on 22 September 2021).
50. Niewiesk, S. (2014). Maternal Antibodies: Clinical Significance, Mechanism of Interference with Immune Responses, and Possible Vaccination Strategies. *Frontiers in Immunology* 5:446. https://doi: 10.3389/fimmu.2014.00446 (Accessed on 6 December 2021).
51. Palm, A.-K. E. & Henry, C. (2019). Remembrance of Things Past: Long-Term B Cell Memory After Infection and Vaccination. *Frontiers in Immunology* 10: 1787. https://doi: 10.3389/fimmu.2019.01787 (Accessed on 27 December 2021).
52. Perdomo-Celis, F., Taborda, N.A. & Rugeles, M.T. (2019). CD8$^+$ T-Cell Response to HIV Infection in the Era of Antiretroviral Therapy. *Frontiers in Immunology* 10: 1896. https://doi: 10.3389/fimmu.2019.01896 (Accessed on 23 December 2021).
53. Qiu, T., Xiao, H., Zhang, Q., Qiu, J., Yang, Y., Wu, D., Cao, Z. & Zhu, R. (2015). Proteochemometric Modeling of the Antigen-Antibody Interaction: New Fingerprints for Antigen, Antibody and Epitope-Paratope Interaction. *PLoS ONE* 10(4): e0122416. https://doi.org/10.1371/journal.pone.0122416 (Accessed on 11 November 2021).
54. Raphael, I., Joern, R.R. & Forsthuber, T.G. (2020). Memory CD4 T

Cells in Immunity and Autoimmune Diseases. *Cells* 9(3): 531. https://doi.org/10.3390/cells9030531 (Accessed on 5 January 2021).

55. Ratajczak, W., Niedź wiedzka-Rystwej, P., Tokarz-Deptula, B. & Deptula, W. (2018). Immunological memory cells. *Central European Journal of Immunology* 43(2): 194-203. https://doi: 10.5114/ceji.2018.77390 (Accessed on 23 December 2021).
56. Reverberi, R. & Reverberi, L. (2007). Factors affecting the antigen-antibody reaction. *Blood Transfusion* 5(4): 227-240. https://doi: 10.2450/2007.0047-07 (Accessed on 26 October 2021).
57. Rodgers, J.R. & Rich, R.R. (2008). Antigens and antigen processing. In: *Clinical Immunology: Principles and Practice (3rd edn.)*; Rich, R.R., Fleisher, T.A., Shearer, W.T., Schroeder, H.W., Frew, A.J. & Weyand, C.M. (Ed.). Philadelphia, Pennsylvania: Elsevier Limited. pp.91-101. https://doi.org/10.1016/B978-0-323-04404-2.10006-5 (Accessed on 17 November 2021).
58. Roghanian, A. & Newman, R. (2021). B Cells. *British Society for Immunology*. https://www.immunology.org/public-information/bitesized-immunology/cells/b-cells (Accessed on 31 October 2021).
59. Rosati, E., Dowds, C.M., Liaskou, E., Henriksen, E.K.K., Karlsen, T.H. & Franke, A. (2017). Overview of methodologies for T-cell receptor repertoire analysis. *BMC Biotechnology* 17: 61. https://doi.org/10.1186/s12896-017-0379-9 (Accessed on 16 December 2021).
60. Rossenspire, A. & Stemmer, P. (2010). Antigen-Specific Signal Transduction. In: *Comprehensive Toxicology (2nd edn.)*; McQueen, C.A. (Ed.). Philadelphia, Pennsylvania: Elsevier Limited. pp.323-349. https://doi.org/10.1016/B978-0-08-046884-6.00619-9 (Accessed on 24 November 2021).
61. Schroeder, H.W. & Cavacini, L. (2010). Structure and function of immunoglobulins. *The Journal of Allergy and Clinical Immunology* 125(2): S41-S52. https://doi.org/10.1016/j.jaci.2009.09.046 (Accessed on 8 September 2021).
62. Schroeder, H.W., Jr., Imboden, J.B. & Torres, R. M. (2019). Antigen Receptor Genes, Gene Products and Coreceptors. In: *Clinicla Immunology: Principles and Practice (5th edn.)*; Rich, R.R., Fleisher, T.A., Shearer, W.T., Schroeder, H.W., Jr., Frew, A.J. & Weyand, C.A. (Eds.). Philadelphia, Pennsylvania: Elsevier Limited. pp.55-77.e1. https://doi.org/10.1016/B978-0-7020-6896-6.00004-1 (Accessed on 16 December 2021).

63. Seder, R.A. % Ahmed, R. (2003). Similarities and differences in CD4 and CD8 effector and memory T cell generation. *Nature Immunology* 4: 835-842. https://doi.org/10.1038/ni969 (Accessed on 31 December 2021).
64. Sela-Culang, I., Kunik, V. & Ofan, Y. (2013). The structural basis of antibody-antigen recognition. *Frontiers in Immunology* 4: 302. https://doi: 10.3389/fimmu.2013.00302 (Accessed on 25 October 2021).
65. Sutton, B.J., Davies, A.M., Bax, H.J. & Karagiannis, S.N. (2019). IgE Antibodies: From Structure to Function and Clinical Translation. *Antibodies* 8(1): 19. https://doi.org/10.3390/antib8010019 (Accessed on 29 September 2021).
66. Takeuchi, A. & Saito, T. (2017). CD4 CTL, a Cytotoxic Subset of $CD4^+$ T Cells, Their Differentiation and Function. *Frontiers in Immunology* 8: 194. https://doi: 10.3389/fimmu.2017.00194 (Accessed on 22 December 2021).
67. ThermoFisher Scientific (2021). *Introduction to Immunoglobulins*. https://www.thermofisher.com/ph/en/home/life-science/antibodies/antibodies-learning-center/antibodies-resource-library/antibody-methods/introduction-immunoglobulins.html (Accessed on 20 September 2021).
68. Turner, D.L. & Farber, D.L. (2014). Mucosal resident memory CD4 T cells in protection and immunopathology. *Frontiers in Immunology* 5:331. https://doi: 10.3389/fimmu.2014.00331 (Accessed on 6 January 2021).
69. Vos, Q., Lees, A., Wu, Q., Snapper, C.M. & Mond, J.J. (2000). B-cell activation by the T-cell-independent type 2 antigens as an integral part of the humoral immune response to pathogenic microorganisms. *Immunological Reviews* 176: 154-170. https://doi: 10.1034/j.1600-065x.2000.00607.x. (Accessed on 23 November 2021).
70. Warrington, R., Watson, W., Kim, H.L. & Antonetti, F.R. (2011). An introduction to immunology and immunopathology. *Allergy, Asthma & Clinical Immunology* 7: S1. https://doi.org/10.1186/1710-1492-7-S1-S1 (Accessed on 11 November 2021).
71. Wissinger, E. (2021). CD8+ T Cells. *British Society for Immunology*. https://www.immunology.org/public-information/bitesized-immunology/cells/cd8-t-cells (Accessed on 22 December 2021).
72. Zhang, N. & Bevan, M.J. (2011). $CD8^+$ T Cells: Foot Soldiers of the

Immune System. *Immunity* 35(2): 161-168. https://doi: 10.1016/j.immuni.2011.07.010 (Accessed on 22 December 2021).

FIGURE REFERENCES

73. Opening image. National Institutes of Health, 2015. *Public domain*; https://commons.wikimedia.org/wiki/File:Killer_T_cells_surround_a_cancer_cell.png (Accessed on 8 September 2021).
74. Figure 3.1. Sjef (2008). *Creative commons attribution-sharealike 3.0 unported license*; https://commons.wikimedia.org/wiki/File:Antigen_presentation.svg (Accessed on 28 October 2021).
75. Figure 3.2. Sjef (2008). *Creative commons attribution-sharealike 3.0 unported license*; https://commons.wikimedia.org/wiki/File:Antigen_presentation.svg (Accessed on 28 October 2021).
76. Figure 3.3. Bobologist (2015). *Creative commons attribution-sharealike 4.0 international license*; https://commons.wikimedia.org/wiki/File:Transitional_B_cell_development.PNG (Accessed on 31 October 2021).
77. Figure 3.4. Bobologist (2015). *Creative commons attribution-sharealike 3.0 unported license*; https://commons.wikimedia.org/wiki/File:B_cell_activation_naive_to_plasma_cell.png (Accessed on 31 October 2021).
78. Figure 3.5. Kovaltsuk, A., Krawczyk, K., Galson, J.D., Kelly, D.F., Deane, C.M. & Trück, J. (2017). How B-Cell Receptor Repertoire Sequencing Can Be Enriched with Structural Antibody Data. *Frontiers in Immunology* 8: 1753. *Creative commons attribution 4.0 international license*;https://doi: 10.3389/fimmu.2017.01753 (Accessed on 9 September 2021).
79. Figure 3.6. (Upper image) Jones, K., Savulescu, A.F., Brombacher, F. & Hadebe, S. (2020). Immunoglobulin M in Health and Diseases: How Far Have We Come and What Next? *Frontiers in Immunology* 11: 595535. *Creative commons attribution 4.0 international license*; https://doi: 10.3389/fimmu.2020.595535 (Accessed on 23 September 2021). (Lower images) Keyt, B.A., Baliga, R., Sinclair, A.M., Carroll, S.F. & Peterson, M.S. (2020). Structure, Function, and Therapeutic Use of IgM Antibodies. *Antibodies* 9(4): 53. *Creative commons attribution 4.0 international license*;https://doi.org/10.3390/antib9040053 (Accessed on 22 September 2021).

80. Figure 3.7. Keyt, B.A., Baliga, R., Sinclair, A.M., Carroll, S.F. & Peterson, M.S. (2020). Structure, Function, and Therapeutic Use of IgM Antibodies. *Antibodies* 9(4): 53. *Creative commons attribution 4.0 international license*; https://doi.org/10.3390/antib9040053 (Accessed on 22 September 2021).
81. Figure 3.8. Keyt, B.A., Baliga, R., Sinclair, A.M., Carroll, S.F. & Peterson, M.S. (2020). Structure, Function, and Therapeutic Use of IgM Antibodies. *Antibodies* 9(4): 53. *Creative commons attribution 4.0 international license*;https://doi.org/10.3390/antib9040053 (Accessed on 22 September 2021).
82. Figure 3.9. de Taeye, S.W., Rispens, T. & Vidarsson, G. (2019). The Ligands for Human IgG and Their Effector Functions. *Antibodies* 8(2): 30. *Creative commons attribution 4.0 international license*;https://doi.org/10.3390/antib8020030 (Accessed on 23 September 2021).
83. Figure 3.10. Breedveld, A. & van Egmond, M. (2019). IgA and FcαRI: Pathological Roles and Therapeutic Opportunities. *Frontiers in Immunology* 10: 553. *Creative commons attribution 4.0 international license*;https://doi: 10.3389/fimmu.2019.00553 (Accessed 28 September 2021).
84. Figure 3.11. de Sousa-Pereira, P. & Woof, J.M. (2019). IgA: Structure, Function, and Developability. *Antibodies* 8(4): 57.*Creative commons attribution 4.0 international license*;https://doi.org/10.3390/antib8040057 (Accessed on 24 September 2021).
85. Figure 3.12. Sutton, B.J., Davies, A.M., Bax, H.J. & Karagiannis, S.N. (2019). IgE Antibodies: From Structure to Function and Clinical Translation. *Antibodies* 8(1): 19. *Creative commons attribution 4.0 international license*;https://doi.org/10.3390/antib8010019 (Accessed on 29 September 2021).
86. Figure 3.13. Sutton, B.J., Davies, A.M., Bax, H.J. & Karagiannis, S.N. (2019). IgE Antibodies: From Structure to Function and Clinical Translation. *Antibodies* 8(1): 19. *Creative commons attribution 4.0 international license*;https://doi.org/10.3390/antib8010019 (Accessed on 29 September 2021).
87. Figure 3.14. Backhaus, O. (2018). Generation of Antibody Diversity. In: *Antibody Engineering*; Thomas Böldicke (Ed.). IntechOpen. *Creative commons attribution 3.0 international license*;https://doi:10.5772/intechopen.72818 (Accessed on 11 October 2021).

88. Figure 3.15. Backhaus, O. (2018). Generation of Antibody Diversity. In: *Antibody Engineering*; Thomas Böldicke (Ed.). IntechOpen. *Creative commons attribution 3.0 international license*;https://doi:10.5772/intechopen.72818 (Accessed on 11 October 2021).
89. Figure 3.16. Backhaus, O. (2018). Generation of Antibody Diversity. In: *Antibody Engineering*; Thomas Böldicke (Ed.). IntechOpen. *Creative commons attribution 3.0 international license*;https://doi:10.5772/intechopen.72818 (Accessed on 11 October 2021).
90. Figure 3.17. Backhaus, O. (2018). Generation of Antibody Diversity. In: *Antibody Engineering*; Thomas Böldicke (Ed.). IntechOpen. *Creative commons attribution 3.0 international license*;https://doi:10.5772/intechopen.72818 (Accessed on 11 October 2021).
91. Figure 3.18. Backhaus, O. (2018). Generation of Antibody Diversity. In: *Antibody Engineering*; Thomas Böldicke (Ed.). IntechOpen. *Creative commons attribution 3.0 international license*;https://doi:10.5772/intechopen.72818 (Accessed on 11 October 2021).
92. Figure 3.19. Norman. R.A., Ambrosetti, F., Bonvin, A.M.J., Colwell, L.J., Kelm, S., Kumar, S. & Krawczyk, K. (2019). Computational approaches to the therapeutic antibody design: Established methods and emerging trends. *Briefings in Bioinformatics* 21(5). *Creative commons attribution 4.0 international license*; https://doi:10.1093/bib/bbz095 (Accessed on 15 November 2021).
93. Figure 3.20. OpenStax, 2012. Microbiology. *Creative commons attribution 4.0 international license*; https://opentextbc.ca/microbiologyopenstax/chapter/overview-of-specific-adaptive-immunity/ (Accessed on 1 December 2021).
94. Figure 3.21. Sela-Culang, I., Kunik, V. & Ofan, Y. (2013). The structural basis of antibody-antigen recognition. *Frontiers in Immunology* 4: 302. *Creative commons attribution 3.0 unported license*;https://doi:10.3389/fimmu.2013.00302 (Accessed on 25 October 2021).
95. Figure 3.22. Ledsgaard, L., Jenkins, T.P., Davidsen, K., Krause, K.E., Martos-Esteban, A., Engmark, M., Andersen, M.R., Lund, O. & Laustsen, A.H. (2018). Antibody Cross-Reactivity in Antivenom Research. *Toxins* 10(10): 393. *Creative commons attribution 4.0 international license*; https://doi.org/10.3390/toxins10100393 (Accessed on 15 November 2021).

96. Figure 3.23. Altaileopard, (2010). *Public domain*; https://commons.wikimedia.org/wiki/File:T-dependent_B_cell_activation.png (Accessed on 27 November 2021).
97. Figure 3.24. Molnar, C. & Gair, J. (2021). Concepts of Biology-1st Canadian Edition. *Creative commons attribution 4.0 international license*; https://opentextbc.ca/biology/chapter/23-2-adaptive-immune-response/ (Accessed on 1 December 2021).
98. Figure 3.25. Hotinger, J.A. & May, A.E. (2020). Antibodies Inhibiting the Type III Secretion System of Gram-Negative Pathogenic Bacteria. *Antibodies* 9(3): 35.*Creative commons attribution 4.0 international license*;https://doi.org/10.3390/antib9030035 (Accessed on 1 December 2021).
99. Figure 3.26. Motran, C.C., Silvane, L., Chiapello, L.S., Theumer, M.G., Ambrosio, L.F., Volpini, X., Celias, D.P. & Cervi, L. (2018). Helminth Infections: Recognition and Modulation of the Immune Response by Innate Immune Cells. *Frontiers in Immunology* 9: 664. *Creative commons attribution 4.0 international license*; https://doi: 10.3389/fimmu.2018.00664 (Accessed on 2 December 2021).
100. Figure 3.27. Albrecht, M. & Arck, P.C. (2020). Vertically Transferred Immunity in Neonates: Mothers, Mechanisms and Mediators. *Frontiers in Immunology* 11:555. *Creative commons attribution 4.0 international license*; https://doi.org/10.3389/fimmu.2020.00555 (Accessed on 6 December 2021).
101. Figure 3.28. Albrecht, M. & Arck, P.C. (2020). Vertically Transferred Immunity in Neonates: Mothers, Mechanisms and Mediators. *Frontiers in Immunology* 11:555. *Creative commons attribution 4.0 international license*; https://doi.org/10.3389/fimmu.2020.00555 (Accessed on 6 December 2021).
102. Figure 3.29. Betts, J.G., Young, K.A., Wise, J.A., Johnson, E., Poe, B., Kruse, D.H., Korol, O., Johnson, J.E., Womble, M. & DeSaix, P. (2013). Anatomy and Physiology. OpenStax: Houston, Texas. *Creative commons attribution 4.0 license*; https://openstax.org/books/anatomy-and-physiology/pages/21-6-diseases-associated-with-depressed-or-overactive-immune-responses (Accessed on 8 December 2021).
103. Figure 3.30. Rosati, E., Dowds, C.M., Liaskou, E., Henriksen, E.K.K., Karlsen, T.H. & Franke, A. (2017). Overview of methodologies for T-cell receptor repertoire analysis. *BMC Biotechnology* 17: 61. *Creative*

commons attribution 4.0 international license; https://doi.org/10.1186/s12896-017-0379-9 (Accessed on 16 December 2021).

104. Figure 3.31. Kumaresan, P.R., da Silva, T.A. & Kontoyiannis, D.P. (2018). Methods of Controlling Invasive Fungal Infections Using CD8$^+$T Cells. *Frontiers in Immunology* 8:1939. *Creative commons attribution 4.0 international license*; https://doi: 10.3389/fimmu.2017.01939 (Accessed on 17 December 2021).
105. Figure 3.32. Tau, Y., Wang, Q., Korner, H., Zhang, L. & Wei, W. (2018). Molecular Mechanisms of T Cells Activation by Dendritic Cells in Autoimmune Diseases. *Frontiers in Pharmacology* 9: 642. *Creative commons attribution 4.0 international license*; https://doi: 10.3389/fphar.2018.00642 (Accessed on 17 December 2021).
106. Figure 3.33. Chandler, N.J., Call, M.J. & Call, M.E. (2020). T Cell Activation Machinery: Form and Function in Natural and Engineered Immune Receptors. *International Journal of Molecular Sciences* 21(19): 7424. *Creative commons attribution 4.0 international license*; https://doi.org/10.3390/ijms21197424 (Accessed on 17 December 2021).
107. Figure 3.34. O'Donnell, H. & McSorley, S.J. (2014). *Salmonella* as aa model for non-cognate Th1 cell stimulation. *Frontiers in Immunology* 5: 621. *Creative commons attribution 4.0 international license*;https://doi: 10.3389/fimmu.2014.00621 (Accessed on 22 December 2021).
108. Figure 3.35. Golubovskaya, V. & Wu, L. (2016). Different Subsets of T Cells, Memory, Effector Functions, and CAR-T Immunotherapy. *Cancers (Basel)* 8(3): 36. *Creative commons attribution 4.0 international license*; https://doi: 10.3390/cancers8030036 (Accessed on 3 January 2022).
109. Figure 3.36. Schluck, M., Hammink, R., Figdor, C.G., Verdoes, M. & Weiden, J. (2019). Biomaterial-Based Activation and Expansion of Tumor-Specific T Cells. *Frontiers in Immunology* 10: 931. *Creative commons attribution 4.0 international license*;https://doi: 10.3389/fimmu.2019.00931 (Accessed on 22 December 2021). Inset: Busselaar, J., Tian, S., van Eenennaam, H. & Borst, J. (2020). Helpless Priming Sends CD8$^+$ T Cells on the Road to Exhaustion. *Frontiers in Immunology* 11: 592569. *Creative commons attribution 4.0 international license*; https://doi: 10.3389/fimmu.2020.592569 (Accessed on 22 December 2021).

110. Figure 3.37. O'Donnell, H. & McSorley, S.J. (2014). *Salmonella* as aa model for non-cognate Th1 cell stimulation. *Frontiers in Immunology* 5: 621. *Creative commons attribution 4.0 international license*;https://doi: 10.3389/fimmu.2014.00621 (Accessed on 22 December 2021).

111. Figure 3.38. Perdomo-Celis, F., Taborda, N.A. & Rugeles, M.T. (2019). CD8+ T-Cell Response to HIV Infection in the Era of Antiretroviral Therapy. *Frontiers in Immunology* 10: 1896. *Creative commons attribution 4.0 international license*;https://doi: 10.3389/fimmu.2019.01896 (Accessed on 23 December 2021).

112. Figure 3.39. Palm, A.-K. E. & Henry, C. (2019). Remembrance of Things Past: Long-Term B Cell Memory After Infection and Vaccination. *Frontiers in Immunology* 10: 1787. *Creative commons attribution 4.0 international license*;https://doi: 10.3389/fimmu.2019.01787 (Accessed on 27 December 2021).

113. Figure 3.40. Palm, A.-K. E. & Henry, C. (2019). Remembrance of Things Past: Long-Term B Cell Memory After Infection and Vaccination. *Frontiers in Immunology* 10: 1787. *Creative commons attribution 4.0 international license*;https://doi: 10.3389/fimmu.2019.01787 (Accessed on 27 December 2021).

114. Figure 3.41. Jaigirdar, S.A. & MacLeod, M.K.L. (2015). Development and function of protective and pathologic memory CD4 T cells. *Frontiers in Immunology* 6: 456. *Creative commons attribution 4.0 international license*;https://doi: 10.3389/fimmu.2015.00456 (Accessed on 6 January 2022).

115. Figure 3.42. Raphael, I., Joern, R.R. & Forsthuber, T.G. (2020). Memory CD4 T Cells in Immunity and Autoimmune Diseases. *Cells* 9(3): 531. *Creative commons attribution 4.0 international license*;https://doi.org/10.3390/cells9030531 (Accessed on 5 January 2021).

116. Figure 3.43. Golubovskaya, V. & Wu, L. (2016). Different Subsets of T Cells, Memory, Effector Functions, and CAR-T Immunotherapy. *Cancers (Basel)* 8(3): 36. *Creative commons attribution 4.0 international license*; https://doi: 10.3390/cancers8030036 (Accessed on 3 January 2022).

117. Figure 3.44. Golubovskaya, V. & Wu, L. (2016). Different Subsets of T Cells, Memory, Effector Functions, and CAR-T Immunotherapy. *Cancers (Basel)* 8(3): 36. *Creative commons attribution 4.0 international license*; https://doi: 10.3390/cancers8030036 (Accessed on 3 January 2022).

118. Figure 3.45. Raphael, I., Joern, R.R. & Forsthuber, T.G. (2020). Memory CD4 T Cells in Immunity and Autoimmune Diseases. *Cells* 9(3): 531. *Creative commons attribution 4.0 international license*;https://doi.org/10.3390/cells9030531 (Accessed on 5 January 2021).

119. Figure 3.46. Raphael, I., Joern, R.R. & Forsthuber, T.G. (2020). Memory CD4 T Cells in Immunity and Autoimmune Diseases. *Cells* 9(3): 531. *Creative commons attribution 4.0 international license*;https://doi.org/10.3390/cells9030531 (Accessed on 5 January 2021).

120. Figure 3.47. Turner, D.L. & Farber, D.L. (2014). Mucosal resident memory CD4 T cells in protection and immunopathology. *Frontiers in Immunology* 5:331. *Creative commons attribution 3.0 unported license* ; https://doi: 10.3389/fimmu.2014.00331 (Accessed on 6 January 2021).

INDEX

D

E

F

G

N

O

P

R